Soft Computing Applications in Sensor Networks

Soft Computing Applications in Sensor Networks

Edited by
Sudip Misra
Sankar K. Pal

CRC Press
Taylor & Francis Group
Boca Raton   London   New York

CRC Press is an imprint of the
Taylor & Francis Group, an **informa** business
A CHAPMAN & HALL BOOK

CRC Press
Taylor & Francis Group
6000 Broken Sound Parkway NW, Suite 300
Boca Raton, FL 33487-2742

© 2017 by Taylor & Francis Group, LLC
CRC Press is an imprint of Taylor & Francis Group, an Informa business

Library of Congress Cataloging-in-Publication Data

Names: Misra, Sudip, editor. | Pal, Sankar K., editor
Title: Soft computing applications in sensor networks / editors, Sudip Misra.
and Sankar K. Pal.
Description: Boca Raton : Taylor & Francis, a CRC title, part of the Taylor &
Francis imprint, a member of the Taylor & Francis Group, the academic
division of T&F Informa, plc, [2017] | Includes bibliographical references
and index.
Identifiers: LCCN 2016008478 | ISBN 9781482298758 (alk. paper)
Subjects: LCSH: Wireless sensor networks--Computer programs. | Soft computing.
Classification: LCC TK7872.D48 S635 2017 | DDC 004.6--dc23
LC record available at https://lccn.loc.gov/2016008478

Visit the Taylor & Francis Web site at
http://www.taylorandfrancis.com

and the CRC Press Web site at
http://www.crcpress.com

Printed and bound in the United States of America by Publishers Graphics,
LLC on sustainably sourced paper.

*Dedicated to
Our Families*

Contents

III   Advanced Topics   171

Preface

Overview and Goals

Soft computing is an emerging methodology for robust and cost-effective problem solving involving uncertain environments. The different features of soft computing facilitate real-time information processing in the presence of such environments. Thus, several solution techniques, such as fuzzy logic, ant colony optimization, particle swarm optimization, and genetic algorithms, have major impacts in solving problems where precise mathematical models are unavailable.

Concurrently, sensor networking is popular for real-time monitoring involving remote sensing, ubiquitous health monitoring, target tracking, and localization. A sensor network is an integration of typically heterogeneous sensors (hardware) with wireless communication infrastructure, middleware, and software tools. The general problem can be broken down into several steps, including channel access, routing, data aggregation, location estimation, and target tracking, while sensor nodes are known as energy constraint devices.

Recently, soft computing methodologies, such as fuzzy logic and ant colony optimization, are demonstrated to be promising for solving different problems in sensor networks such as efficient learning and uncertainty handling. Several conference and journal papers on applying soft computing techniques in sensor networking have appeared in the past few years. They describe various challenging issues related to sensor networking using soft computing techniques. Moreover, different researchers address the problems from different perspectives. Therefore, there has been a need for a book describing in a consolidated manner the major recent trends of soft computing to provide comprehensive information to researchers, applied scientists, and practitioners.

This handbook is written by worldwide experts, with the aim of bringing together research works describing soft computing approaches in sensor networking, while investigating the novel solutions and discussing the future trends in this field. It includes tutorials and new material that describe basic concepts, theory, and algorithms that demonstrate why and how soft computing techniques can be used for sensor networking in different disciplines. All the chapters provide a balanced mixture of methodologies and applications. After a brief tutorial-style introduction, each chapter contains a comprehensive description of the developments in its respective area, and is written to blend well with the other chapters.

This book is useful to graduate students, researchers, and practitioners working in different fields spanning computer science, system science, and information technology, as a reference book and as a text book.

Organization and Features

The book is broadly divided into three parts. Part I consists of two chapters. Chapter 1 presents preliminary concepts of sensor networks. Chapter 2 is dedicated to the recent advances in soft computing. Part II, comprised of five chapters, focuses on the recent advances in soft computing applications in sensor networks. Chapter 3 discusses the evolution of soft computing in sensor networks in different application scenarios. Chapters 4 and 5 discuss routing mechanisms in sensor networks using soft computing applications. Chapter 6 describes game theoretic aspects in wireless sensor networks.

Chapter 7 is dedicated to energy efficiency and bounded hop data delivery in sensor networks using a multi-objective optimization approach. Part III consists of four chapters, and is dedicated to the advanced topics in sensor networks in which different problems can be potentially solved using soft computing applications. Chapter 8 discusses context aware services in sensor networks. Chapter 9 presents energy-aware wireless body area networks (WBANs) for critical health care information delivery, while presenting different aspects of soft computing applications in WBANs. Chapter 10 is dedicated to the complex network entropy in the context of sensor networks. Finally, Chapter 11 discusses challenges and possibilities of soft computing approaches which are applicable for ad hoc sensor networks specifically for vehicular networks (VANETs).

We list some of the important features of this book, which, we believe make this book a valuable resource to our readers.

- Most chapters are written by prominent academicians, researchers, and practitioners working in respective topical areas for several years and have thorough understanding of the concepts.

- Most chapters focus on future research directions and target researchers working in these areas. They provide insight to researchers about some of the current research issues.

- The authors represent diverse nationalities. This diversity enriches the book by including ideas from different parts of the world.

- At the end of each chapter, we included additional literature, which readers can use to enhance their knowledge.

Target Audience

This book is written for the student community at all levels from those new to the field to undergraduates and postgraduates. It is designed to be useful at all learning levels.

The secondary audiences for this book are the research communities in both academia and industry. To meet the specific needs to these groups, most chapters include sections discussing directions of future research.

Finally, we have considered the needs of readers from the industries who seek practical insights, for example, how soft computing applications are useful in real-life sensor networks.

List of Figures

List of Tables

Editors

Dr. Sudip Misra is an Associate Professor in the School of Information Technology at the Indian Institute of Technology Kharagpur. He earned a Ph.D. in Computer Science at Carleton University, Ottawa, Canada. His current research interests include algorithm design for emerging communication networks. He is the author of over 170 scholarly research papers, more than 80 of which were published by reputable organizations such as IEEE, ACM, Elsevier, Wiley, and Springer.

Dr. Misra's work was recognized at various conferences. He received the IEEE ComSoc Asia Pacific Outstanding Young Researcher Award in 2012. He was also the recipient of several academic awards and fellowships including the Young Scientist Award of the National Academy of Sciences of India, the Young Systems Scientist Award of the Systems Society of India, the Young Engineers Award of the Institution of Engineers, Carleton University's Governor General's Academic Gold Medal awarded to an outstanding graduate student doctoral level, and the Swarna Jayanti Puraskar Golden Jubilee Award of the National Academy of Sciences of India. Dr. Misra also received a prestigious NSERC post-doctoral fellowship from the Canadian government and a Humboldt research fellowship in Germany.

Dr. Misra is the editor-in-chief of the *International Journal of Communication Networks and Distributed Systems* (Interscience, UK), and an associate editor of the *Telecommunication Systems Journal* (Springer), *International Journal of Communication Systems* (Wiley), and EURASIP's *Journal of Wireless Communications and Networking*. He also serves as an editor, editorial board member, or editorial review board member for the IET's *Communications Journal* and *Wireless Sensor Systems Journal* and the *Computers and Electrical Engineering Journal* (Elsevier).

Dr. Misra is writing two books and edited seven books published by well known companies such as Springer, Wiley, Cambridge University Press, and World Scientific. He has chaired several international conference and workshops and was invited to deliver keynote lectures at more than 20 conferences in the US, Canada, Europe, Asia, and Africa.

Professor Sankar K. Pal (www.isical.ac.in/~sankar) is a distinguished scientist of the Indian Statistical Institute and its former director. He is also a J.C. Bose Fellow of the Government of India and former chair professor of the Indian National Academy of Engineering. He founded the Machine Intelligence Unit and the Center for Soft Computing Research, a national facility in Calcutta. Professor Pal earned a Ph.D. in Radio Physics and Electronics from the University of Calcutta in 1979, and another Ph.D. and DIC in Electrical Engineering from the Imperial College, University of London in 1982. He joined the Indian Statistical Institute in 1975 and advanced to full professor in 1987 and distinguished scientist in 1998, and served as its director from 2005 to 2010.

Professor Pal worked at the University of California Berkeley, the University of Maryland College Park, NASA Johnson Space Center, and the U.S. Naval Research Laboratory. Since 1997, he has served as a distinguished visitor for the Asia-Pacific Region on behalf of the IEEE Computer Society and held visiting positions at several universities in Italy, Poland, Hong Kong, and Australia.

Professor Pal is a Fellow of the IEEE, the Academy of Sciences for the Developing World, the International Association for Pattern Recognition, the International Association of Fuzzy Systems, the International Rough Set Society, and four National Academies for Science and Engineering in India. He is a co-author of 17 books and more than 400 research publications in the areas of pattern recognition and machine learning, image processing, data mining and Web intelligence, soft computing, neural nets, genetic algorithms, fuzzy and rough sets, and bioinformatics.

Among his numerous awards granted by Indian and international governments and societies, Dr. Pal received the 1990 S.S. Bhatnagar Prize (the most coveted award for a scientist in India), the 2013 Padma Shri (one of the highest civilian awards) by the President of India, the G.D. Birla Award, the Om Bhasin Award, a Jawaharlal Nehru Fellowship, the Khwarizmi International Award from the President of Iran, the FICCI Award, the Vikram Sarabhai Research Award, NASA Tech Brief Award, IEEE Transactions on Neural Networks Outstanding Paper Award, a NASA Patent Application Award, IETE-R.L. Wadhwa Gold Medal, INSA-S.H. Zaheer Medal, the Indian Science Congress's P.C. Mahalanobis Birth Centenary Gold Medal for lifetime achievement, the J.C. Bose Fellowship of the Government of India, and an INAE chair professorship.

Professor Pal has filled various editorial positions and served on editorial boards throughout his career. He has been associated with *IEEE Transactions on Pattern Analysis and Machine Intelligence, IEEE Transactions on Neural Networks, Neurocomputing, Pattern Recognition Letters*, the *International Journal of Pattern Recognition and Articial Intelligence, Applied Intelligence, Information Sciences, Fuzzy Sets and Systems, Fundamenta Informaticae, LNCS Transactions on Rough Sets*, the *International Journal of Computational Intelligence and Applications, IET Image Processing*, the *Journal of Intelligent Information Systems*, and the *International Journal of Signal Processing, Image Processing and Pattern Recognition*. He served as book series editor for Frontiers in Artificial Intelligence and Applications (IOS Press) and Statistical Science and Interdisciplinary Research (World Scientific); as a member of the advisory editorial boards of *IEEE Transactions on Fuzzy Systems*, the *International Journal on Image and Graphics*, and the *International Journal of Approximate Reasoning*. He also served as a guest editor for several journals including *IEEE Computer*, *IEEE T-SMC*, and *Theoretical Computer Science*.

Contributors

Angelos Antonopoulos
Telecommunications Technological
  Centre of Catalonia
Barcelona, Spain

Seema Bawa
Thapar University
Patiala, India

Samaresh Bera
Indian Institute of Technology
  Kharagpur
Kharagpur, India

Hindol Bhattacharya
Jadavpur University
Kolkata, India

Abhishek Chakraborty
Indian Institute of Space Science and
  Technology
Thiruvananthapuram, India

Sandip Chakraborty
Indian Institute of Technology
  Kharagpur
Kharagpur, India

Matangini Chattopadhyay
Jadavpur University
Kolkata, India

Samiran Chattopadhyay
Jadavpur University
Kolkata, India

Sankar N. Das
Indian Institute of Technology
  Kharagpur
Kharagpur, India

Biswanath Dey
National Institute of Technology
Silchar, India

Amit Dua
Thapar University
Patiala, India

Ernesto Ibarra
University of Barcelona
Barcelona, Spain

Elli Kartsakli
Technical University of Catalonia
Barcelona, Spain

P. Venkata Krishna
Sri Padmavati Mahila University
Tirupati, AP, India

Neeraj Kumar
Thapar University
Patiala, India

B. S. Manoj
Indian Institute of Space Science and
  Technology
Thiruvananthapuram, India

Sukumar Nandi
Indian Institute of Technology
  Guwahati
Guwahati, India

Gaurang Panchal
Indian Institute of Technology
    Kharagpur
Kharagpur, India

Garimella Rammurthy
International Institute of Information
    Technology Gachibowli
Hyderbad, India

Debasis Samanta
Indian Institute of Technology
    Kharagpur
Kharagpur, India

V. Saritha
VIT University
Vellore, India

Priyanka Sharma
International Institute of Information
    Technology Gachibowli
Hyderbad, India

Priti Singh
Indian Institute of Space Science and
    Technology
Thiruvananthapuram, Kerala, India

S. Sivanesan
VIT University
Vellore, India

Munjuluri Srikanth
International Institute of Information
    Technology Gachibowli
Hyderbad, India

Christos Verikoukis
Telecommunications Technological
    Centre of Catalonia
Castelldefels, Barcelona, Spain

Konjeti Viswanadh
International Institute of Information
    Technology Gachibowli
Hyderbad, India

Part I

Introduction

Chapter 1

Introduction to Wireless Sensor Networks

Sankar N. Das

Indian Institute of Technology Kharagpur

Samaresh Bera

Indian Institute of Technology Kharagpur

Sudip Misra

Indian Institute of Technology Kharagpur

Sankar K. Pal

Indian Statistical Institute Kolkata

1.1   Introduction

Wireless sensor networks (WSNs) are relatively new technologies used in different civilian and military applications. Typically, a WSN consists of several (hundreds or thousands) sensor nodes, and each node is dedicated to monitor some predefined physical parameter. The sensor nodes can be activated or deactivated dynamically based on the requirements. Figure 1.1 shows a WSN comprising different sensor nodes, access points (i.e., gateways), base station, and servers. A sensor network consists of homogeneous sensors or heterogeneous sensors. In homogeneous sensors, physical properties of all the sensors are the same and dedicated to monitor a particular task. In contrast, a heterogeneous sensor network consists of different sensor nodes having different physical properties and dedicated to perform different tasks in a region. The sensor nodes can be dedicated to perform entity monitoring, area monitoring, and entity-area monitoring. Additionally, a sensor network can perform different tasks in a hierarchical manner, as shown in Figure 1.1. All sensor nodes are homogeneous in nature and send the sensed information directly to the gateways. Similarly, in case of single-tier and heterogeneous sensor nodes, the sensors are heterogeneous and send the sensed information directly to the gateways. In contrast, in multi-tier sensor networks, the sensed information is sent to the gateways in a hierarchical manner, i.e., from one sub-network to another sub-network and eventually, to the gateways. After collecting the sensed information from the WSN nodes, the gateways relay the information to the base stations based on some *intelligence*. At the server side, the information is computed and processed, and adequate decisions are taken to perform the monitoring task efficiently. The sensor nodes can also communicate with one another to perform the given task in a structured and collaborative manner. *Target tracking* is one of the applications of sensor networks in which a task is performed in a structured and collaborative manner.

1.2   Classification of WSNs

A WSN is categorized into multiple domains based on the attributes of application, transmission media, types of sensor nodes, and type of network as depicted in Figure 1.2. Wireless sensor networks are used in different applications such as military, environment monitoring, health condition monitor-

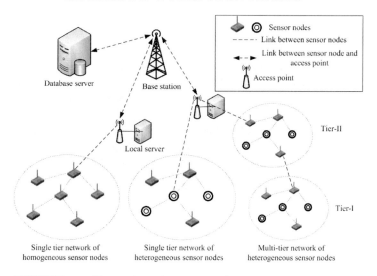

FIGURE 1.1: Network architecture of wireless sensor network

ing, industrial (e.g., monitoring health of a structure), agricultural (e.g., soil moisture and soil temperature), and vehicular networks. Typically, the transmission medium used for communication in a sensor network is either radio frequency or acoustic. On the other hand, sensor nodes can be static or mobile. Additionally, a sensor node can sense multiple things simultaneously. The sensor nodes can perform their tasks in a structured or unstructured manner. Each of the sub-components is shown below.

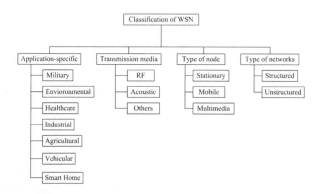

FIGURE 1.2: Classification of wireless sensor networks

1.2.1   Specific Applications

The main objective of the wireless sensor network is to monitor some applications such as temperature, pressure, speed, and detecting fire. Based on the applications, the WSN can be categorized into multiple domains — military applications, environmental, health care, industrial, agricultural, vehicular networks, and smart homes. Initially, the deployment of sensor networks started with the military applications, and then attracted interest from other domains due to its features. Consequently, sensor networks are widely used in different sectors to monitor different tasks efficiently.

1.2.2   Transmission Media: Radio Frequency, Acoustic, and Others

We already mentioned in Section 1.1 that the sensor nodes can communicate with one another to exchange information among them. The major communication medium used in sensor network is radio frequency (RF). The RF communication method is widely used for terrestrial applications. On the other hand, acoustic communication is useful in underwater surveillance systems as RF communications cannot be used in such cases. Some sensor networks use both the RF and acoustic communication modes, where both the terrestrial and underwater surveillance are present. Further, different researchers proposed other types of communication methods in adverse environments, where both the RF and acoustic communication may not work.

1.2.3   Types of Nodes: Static, Mobile, and Multimedia

The sensor nodes in a WSN can be static or mobile or both based on the requirements. In case of static WSNs, the sensor nodes cannot change their positions; they remain fixed according to their initial deployment. On the other hand, sensor nodes are mobile in mobile WSNs. In such WSNs, the sensor nodes can adjust their positions to monitor the dedicated task in an efficient manner. Additionally, both the static and mobile WSNs can be used for multimedia applications. In some cases, the sensor nodes only monitor scalar parameters such as temperature and pressure. However, currently, sensor networks are also widely used to monitor vector parameters such as video surveillance. In such systems, both the scalar and vector sensors are deployed, and they are activated dynamically based on the reported information from the sensor nodes. However, the WSNs are very expensive for multimedia applications rather than for monitoring scalar parameters.

1.2.4   Types of Networks

We see that the sensor networks can monitor different applications as mentioned in Section 1.2.1. Therefore, the sensor nodes are required to perform

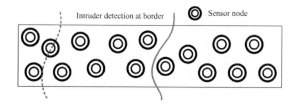

FIGURE 1.3: Intruder detection at border of region

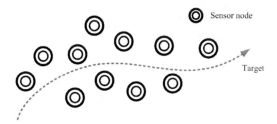

FIGURE 1.4: Tracking target on battlefield

a given task in a structured manner. However, in some cases, the nodes can perform the task in an unstructured manner. In the structured monitoring applications, the sensor nodes need to form a group to perform a particular task, and thereby need some *intelligence* to perform this. On the other hand, the sensor nodes can perform dedicated tasks independently and may not necessarily form a structure.

1.3 Applications of WSNs

1.3.1 Military

Wireless sensor networks are widely used in military applications such as border security monitoring and target tracking. In Figure 1.3, the intruder detection system is presented. The sensor nodes are deployed at the border region to track any intruder who intends to cross the border. On detecting such an intruder, sensor nodes send their information to the nearest base station, and authorized persons can take necessary measures to disrupt any irregular activities. Similarly, sensor networks can be used to track a particular target, a process known as *target tracking*. Figure 1.4 shows an example of target tracking using sensor networks. The sensor nodes are deployed at a region to track different objects. Thus, the sensor nodes detect the movement of the objects and report to the base station. The sensor can also be activated based on the movement of the objects to track them in an efficient manner.

1.3.2   Environmental

Environment monitoring is another important use of sensor networks. Different applications of sensor networks are fire monitoring, CO_2 level monitoring, and wild-animal monitoring. The sensor nodes are deployed in forests and form a network to monitor the above mentioned parameters.

1.3.3   Health Care

Online patient monitoring is an emerging application of sensor networks. Different sensor nodes (such as temperature, pressure, ECG, and glucose) are placed on the patient's body to monitor different parameters. The sensed data are sent to the medical server through gateways. From the server, doctors can retrieve the data and advise the patient to take necessary medicines. In such a system, the patients always do not need to be physically present with their doctors. Thus, a doctor can check multiple patients simultaneously without any problem.

1.3.4   Industrial

Sensor networks are also used in industrial health condition monitoring systems such as analyzing the health of a bridge and building, and chemical percentage in a product. Sensor nodes are deployed at different places on a bridge or in a building to continuously monitor the health condition of these structures. The sensor nodes report to the gateways on detecting any unwanted and malicious activities.

1.3.5   Agriculture

Recently, sensor networks were deployed in agricultural fields to monitor parameters such as soil temperature and moisture. Figure 1.5 depicts an agricultural field with deployment of sensor nodes. Different sensor nodes are deployed on the field and sense the corresponding parameters. The sensed data is forwarded to the gateways and eventually to the distributed servers where the information is processed and computed to enable optimal decisions. According to the processed information, remote users are alerted to take necessary steps (e.g., supply water). Thus, a precision agriculture system can be deployed using wireless sensor networks.

1.3.6   Vehicular Networks

WSNs play a key role to enable intelligent transportation systems (ITS) where vehicles form a network and exchange information. The vehicles upload their collected information to the road side units (RSUs) and download information from the RSUs to get recent updates from the server. The collected informa-

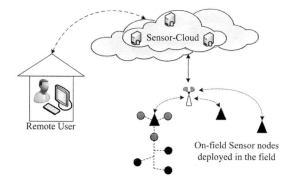

FIGURE 1.5: Wireless sensor network in agriculture

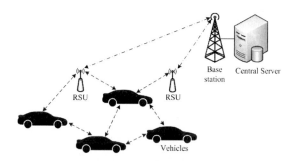

FIGURE 1.6: Vehicular network

tion is transferred to the base station for computation and processing. In Figure 1.6, a vehicular network is presented where vehicles communicate with each other and exchange information to ensure better driving.

1.3.7 Smart Homes

Due to the growing concerns about the energy crisis, people are looking for digitized automated energy supply systems [2]. Sensor networks are widely used in building smart homes, where all appliances are equipped with different sensor nodes to monitor light, temperature and pressure, as shown in Figure 1.7. The sensors automatically control the appliances, and save energy consumption. Thus, cost-effective and reliable energy consumption systems are deployed using the advantages of sensor networks.

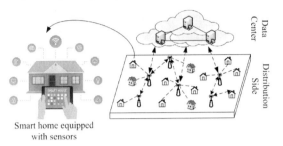

FIGURE 1.7: Smart home equipped with different sensors

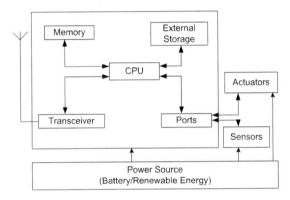

FIGURE 1.8: Sensor node

1.4   Key Factors

1.4.1   System Requirements

Sensor nodes have (i) sensing, (ii) computation, and (iii) communication capabilities. To support such capabilities, the nodes are equipped with (i) microcontroller, (ii) transceiver, (iii) memory, (iv) sensor(s), (v) actuator(s), and (vi) energy sources. A schematic representation is illustrated by Figure 1.8 [8]. One noticeable aspect of the sensor nodes is their resource-constrained nature. Most of the existing hardware platforms or motes, such as Iris [CT] and MICAz [CT], have limited capabilities with respect to communication and computation. However, some high-end sensor platforms also exist, for example, Imote2 [CT] motes. A brief comparison of motes is shown in Table 1.1 [1]. One inadequate resource for sensor nodes is their power source. Most nodes are powered by a pair of AA batteries. However, renewable energy sources now draw much attention from researchers [16].

TABLE 1.1: Mote Hardware Specifications

Mote Type	MCU				OS	Connectivity		
	Processor	Clock Speed (MHz)	Prog. Memory	RAM (kb)		IEEE Standard	Radio Freq.	Trans. Speed (kbps)
Imote2	Intel PXA271 Xscale	13-416	32 MB	256	SOS: linux	IEEE 802.15.4	2.4 GHz	250
Iris	Atmel AT-mega128L	16	128 KB	8	Mote Runner	IEEE 802.15.4	2.4 GHz	250
MICA2	Atmel AT-mega128L	16	128 KB	4	TinyOS	–	433, 868, 916 MHz	38.4 Kbaud
MICAz	Atmel AT-mega128L	16	128 KB	4	Mote Works	IEEE 802.15.4	2.4 GHz	250

1.4.2 Categories of WSNs

WSNs can be categorized on the basis of several factors, such as the deployment environment, types of the nodes, and node mobility. Considering these factors, WSNs are broadly categorized as wireless terrestrial sensor networks, wireless underwater sensor networks, wireless underground sensor networks, wireless multimedia sensor networks, and wireless mobile sensor networks. Next, a short description of each category is provided except for wireless terrestrial sensor networks, the most common WSNs.

Wireless underwater sensor networks: Advancement of wireless underwater sensor networks (WUSNs) generates new opportunities of exploring the flora and fauna of oceanic environments. Moreover, WUSNs are helpful to monitor underwater resources and structures, such as oil rigs. One of the key distinctions of WUSNs is the use of acoustic signal as the mode of communication instead of radio signal [7]. Due to the high attenuation of radio signal in underwater environments, the acoustic signal is a better choice than the radio signal. However, packet communication suffers from large propagation delay, and low link capacity. Other challenging issues are the inherent mobility of the nodes of a WUSN, sparse deployment of costly nodes, and node failure due to environmental conditions [7].

Wireless underground sensor networks: The potential applications of wireless underground sensor networks (WUGNs) includes intelligent agriculture and irrigation, monitoring soil quality, infrastructure, border patrol, and many more [15]. The underground nodes communicate with the aboveground nodes, and both electromagnetic and magnetic induction are used as the communication medium [10]. However, factors, like soil temperature, moisture, composition, and depth affect the quality of communication [10]. The communication range and transmission data rate are smaller with respect to terrestrial WSNs. As an example, the maximum communication range is around 4.5 m when soil moisture is high and the burial depth of the nodes is 35 cm [15].

Wireless multimedia sensor networks: The nodes of a multimedia WSN (WMSN) are equipped with low cost CMOS cameras and microphones [6]. The multimedia nodes communicate the monitored or surveillance data in the form of video, audio, and/or image data. One significant difference of WMSNs is that the multimedia nodes are deployed in a pre-planned manner instead of random deployment for providing target coverage. Generally, the communicated multimedia data is voluminous, and faces variable delay constraints. As a consequence, the nodes of a WMSN have high bandwidth requirement, high energy consumption, and stricter requirements of quality of services.

Wireless mobile sensor networks: Traditionally, WSNs consisted of static nodes that were densely deployed. Those static nodes communicate with the sink through multi-hop communication. Recently, mobile entities used as sinks and also nodes were employed to reduce the communication overhead of the static nodes [4]. Mobility helps the nodes to improve connectivity, reliability, and energy efficiency. On the other hand, mobility introduces some challenges such as mobility management, mobility aware transmission power control, timely detection of mobile entities, and data transfer.

1.4.3   Protocol Stack of WSNs

Successful communication of sensed data is an important issue in WSNs. The protocol stack used for successful communication in WSNs is similar to the traditional TCP/IP protocol stack: application layer, transport layer, network layer, data link layer, and physical layer. However, successful communication depends on other network related issues, such as topology of the network, locations, and transmission power control of the nodes. As the network topology changes over time due to node mobility and node failure, nodes should accommodate those issues at the time of routing of the packets. Moreover, nodes aggregate the data before forwarding to save their energy. A representation of the relationship between the communication protocol and other factors is illustrated by Figure 1.9 [17].

1.4.4   Topology Management

The objective of topology management is to connect the nodes of a WSN in an energy-efficient manner. Topology management can be viewed as utilizing the physical connections or logical relationships among the nodes of a WSN. The resource constrained nodes of WSNs communicate with the sink through multi-hop communication instead of direct communication to save their en-

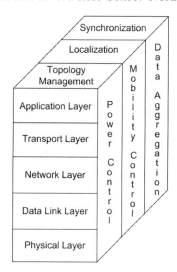

FIGURE 1.9: Protocol stack of WSN

ergy. As a consequence, energy-efficient topology management plays a vital role for enhancing the network efficiency. The proposed approach for topology management for WSNs can be classified as (i) *topology discovery*, (ii) *sleep cycle management*, and (iii) *clustering* [18].

Figure 1.10 illustrates examples of different topologies. For the sake of clarity of the figure, the communication link between a member node and cluster-head is not shown in Figure 1.10(b). A sink or any other node determines the topology of the network by initiating the topology discovery process.

For determining topology, the initiator disseminates a topology discovery request. After receiving such a request, a node may reply with its own information as a response or an aggregated response including the replies from all of its descendants. After receiving the responses from the nodes, the initiator creates and retains a topology map of the network.

One noticeable aspect of the WSNs is that the resource constrained nodes are deployed densely. Those sensor nodes follow an active-sleep cycle to save their energy. By managing the active-sleep cycle properly, the nodes rotate different network related functionalities, such as sensing and relaying data, and also reduce redundancy. Some approaches, e.g., sparse topology and energy management [14], use dual radios, one for data communication, and other, a low energy radio for waking the neighbors.

The sensor nodes also form clusters to reduce the number of the nodes participating in the data transmission process. In clustering approaches, sensor nodes arrange themselves into clusters. A cluster consists of one cluster head and multiple member nodes. The member nodes directly communicate with the cluster head. On the other hand, the cluster heads communicate with the sink directly or by multi-hop communication. The number of clusters is dependent on the total number of nodes, and the size of the clusters may or may not be dependent on the distance between the sink and the cluster heads [28].

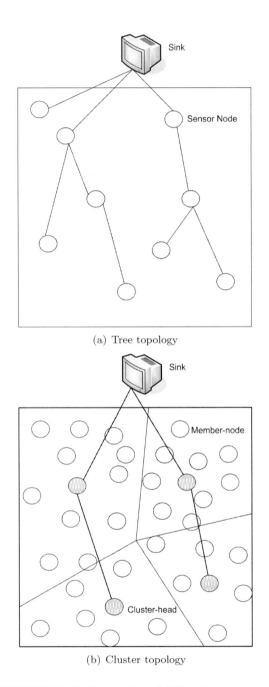

(a) Tree topology

(b) Cluster topology

FIGURE 1.10: Examples of different topologies

1.4.5 Coverage

The coverage of an area of interest by the nodes of WSNs can be specified as the percentage of that area covered or sensed by the nodes. The primary objective of the sensor coverage problem is the estimation of the minimum number of the sensors for providing the full coverage of the area of interest. Another type is k-coverage. If all the crossing points within an area are k-covered, that area is also k-covered. It is assumed that the sensing area of the sensors is a perfect disk, and crossing points are the intersection points between the sensing areas of the neighboring sensor nodes or the intersection points between the sensing areas of the sensor nodes and the boundary of the area. Figure 1.11 illustrates the different types of coverage. The coverage of an area is dependent on the following factors: (i) deployments of nodes, (ii) node mobility, (iii) sensing models of the nodes, (iv) monitored region, and (v) attributes of the application [13].

Sensor nodes may be deployed randomly or deterministically. Moreover, they can be deployed densely or sparsely on the basis of the application-specific and other requirements. The coverage of a WSN is also affected by the mobility of the nodes. For static nodes, the coverage area is fixed. However, the mobile nodes can change the covered area with time and also according to the requirements. Moreover, mobile nodes can enhance the degree of coverage, in case of random deployment, by collaborating among themselves.

As already mentioned and shown by Figure 1.11, the sensed region of the scalar sensors is assumed to be a circular disk. On the other hand, the field of view of the camera sensors is not a disk but a funnel-shaped region in two-dimensional space. The type of the sensing of a node may be binary or probabilistic. In binary model, a node may sense an event or not, i.e., the sensed value is either 1 or 0. On the other hand, the range of the sensed value in the probabilistic model is between 1 and 0. The coverage in WSNs can also be partitioned, depending on the covered area, into the following categories: (i) area coverage, (ii) point coverage, and (iii) barrier coverage [13]. Further, the coverage is also dependent on the application. Target-tracking application can be considered as an example. The coverage parameters of such an application

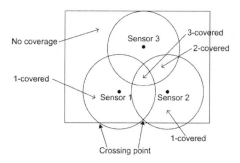

FIGURE 1.11: Different types of coverage within area

are determined by the types of the targets, number of the targets, and velocity of the targets [13].

1.4.6   Important Factors Related to Communication

In this section, a list of the important factors affecting communication is illustrated.

Data Delivery Model: In WSNs, the sensor nodes communicate their sensed data by using one of the following four data delivery models: periodic, event-driven, query-based, and hybrid [11]. In case of periodic data delivery, sensor nodes sense their surrounding environments periodically, and communicate that data to the sink. The presence of an event of interest initiates the communication process in event-driven data delivery. On the other hand, a sink triggers the transmission by the sensors by disseminating a query into the network. In the hybrid data delivery model, nodes follow any combination of the other three data delivery models.

Deployment of Sensor Nodes: Sensor nodes of WSNs can be deployed randomly or deterministically. As an example, nodes are deployed deterministically and linearly in tunnel monitoring applications, whereas nodes of a pollution monitoring application can be deployed randomly within the area of interest. Node deployment affects the network architecture and eventually the communication of the nodes.

Type of Routing: A sensor node may unicast, multicast, broadcast, or anycast its transmitted packets based on the types of the packets, number of sink nodes, requirements of quality of service, and other factors. A node unicasts its packets in the presence of a single sink, or can multicast the same in the presence of multiple sinks. Nodes also broadcast query or control packets. A source node considers different routing strategies on the basis of the location and mobility of the target nodes.

Connectivity: Successful communication between two sensor nodes depends on several factors such as node mobility, link quality, heterogeneity of the nodes, and network dynamism [11]. The communication between two mobile nodes is totally different from the communication among the static ones. As an example, the data collection procedure by the mobile sinks is different from that of the static ones. Link quality is another important factor in communication, and the quality of the links differs due to environmental conditions. The presence of an object, such as a tree or house, affects the link quality, and hence the communication. Network dynamism and the presence of heterogeneous nodes, e.g., a powerful node in terms of communication and battery power, also affect the network architecture and communication between a pair of sensor nodes.

Quality of Service: The requirements of QoS in WSNs are application specific. The requirements of a multimedia WSN or a WSN deployed for health care applications are totally different from those of scalar WSNs. The parameters related to QoS are as follows: priority of the data, end-to-end delay, jitter, throughput, reliability, periodicity of the data, reaction time, and packet loss ratio [3]. Considering the hybrid data delivery model, the priority of the event-driven or query-driven data is more than the data communicated periodically. Event-driven and query-driven data also have higher QoS requirements than the data communicated periodically. The data communicated by multimedia sensor nodes have lower requirements for end-to-end delay and throughput than the scalar sensors. Other parameters have also their own specifications, and are regulated accordingly.

1.4.7 Fault Tolerance

Sensor nodes are prone to failures due to any malfunctioning of the components of the physical, hardware, middleware, system software, and application layer. The fault tolerance approaches designed for WSNs can be partitioned on the basis of their focused layers, i.e., hardware, software, communication, and application. Sensor nodes are equipped with low quality components to remain cost effective. As a consequence, sensor nodes may show faulty behavior due to malfunctioning of low quality hardware components such as memory, battery, microcontroller, sensing units, and communication radio. Moreover, the limited battery power of the nodes poses a constraint on the reliable performance of the nodes.

Another noticeable factor is that the WSNs are generally deployed in harsh environments, and the environmental conditions such as moisture, dust, and heat hamper the performance of the sensor nodes. Further, the sensor nodes may not able to communicate in the presence of any failure of hardware components or link failure. The link failure may be caused by the environmental conditions or interference by the other nodes. On the other hand, software bugs are the most common sources of faults in the software layer. One interesting factor is that the fault tolerance approaches addressed at the application layer are application specific [5].

Generally, there exist two kinds of fault detection approaches: self-diagnosis and cooperative diagnosis. As an example, a node may easily detect failure of any particular link if that link cannot be employed for successful communication within a predetermined time interval. On the other hand, collaboration of neighboring nodes is required to identify a faulty node. In some cases, the fusion center detects the faulty observations or nodes before aggregating the received data. In case of data intensive fault detection mechanisms, various machine learning based approaches, such as support vector machine and Bayesian approaches, are applied [9].

The most common approach for ensuring fault tolerance at network level is redundancy. Multiple source nodes can be activated for fault-tolerant event detection and event monitoring. Multiple disjoint routing paths can be exploited for providing reliability. However, redundancy increases the energy consumption of the nodes, and also the overhead of the protocols. Another way to provide fault tolerance is the use of appropriate fault models. Developing appropriate and realistic fault models is a challenging task because of the types of the sensors, application-specific requirements, and the characteristics of the deployed environment. Heterogeneity, especially at node level, is also employed to ensure fault tolerance [14].

1.4.8   Security

WSNs are susceptible to various types of vulnerabilities and attacks. Generally, WSNs are deployed in open and remote environments and have no or very limited physical security. An adversary can easily capture and tamper with nodes of such WSNs. As the nodes of a WSN are collaborative in nature, a few compromised nodes can affect the other uncompromised nodes, and also disrupt the functionalities of the whole network. Another factor is the inherent broadcast nature of the wireless communication. Adversaries can easily eavesdrop on communicated packets and inject false data or control packets [12].

The primary security concerns for WSNs are protecting communications and avoiding any kind of intrusion. The nodes of WSNs have limited storage, communication, and computational capabilities. As a consequence, the traditional security measures such as public-key encryption are too resource-expensive for WSNs, and are not suitable options. As an example, most of the existing sensor motes, such as MICA motes, do not support frequency hopping or spread spectrum to save their battery power, and they are vulnerable to jamming attack. In WSNs, the security requirements are application specific. Considering that fact and other appropriate context information, e.g., location and time, different levels of security solutions can be provided to mitigate the threats in a resource-efficient manner. A brief list of common attacks in WSNs is catalogued in Table 1.2.

TABLE 1.2: Common Attacks in Wireless Sensor Networks

Layer	Type of Attack	Probable Protection Scheme
Application	Replica attack	Distinct pair-wise keys
Transport	Flooding	Limiting number of connections
Network	Sinkhole, selective forwarding, Sybil attack, HELLO flooding, false routing information	Authentication, redundancy, probing, and encryption
Data Link	Collision, exhaustion	Error control, restriction of communication rate physical and tampering, jamming and efficient key management, tamper-proofing, and spread spectrum scheme

1.5 Concluding Remarks

In this chapter, we discussed different aspects of wireless sensor networks and their different applications. The layered architecture of the sensor networks is also presented. Finally, we discussed hardware platforms which are required to support different sensor network-based applications.

Bibliography

[1] List of wireless sensor nodes, Wikipedia, 2016 [Online].

[2] S. Bera, S. Misra, and J. J. P. C. Rodrigues. Cloud computing applications for smart grid: a survey. *IEEE Transactions on Parallel and Distributed Systems*, 26(5):1477–1494, 2015.

[3] J. Chen, M. Diaz, L. Llopis, and B. Rubio J. M. Troya. A survey on quality of service support in wireless sensor and actor networks: requirements and challenges in the context of critical infrastructure protection. *Journal of Network and Computer Applications*, 34:1225–1239, 2011.

[4] M. D. Francesco, S. K. Das, and G. Anastasi. Data collection in wireless sensor networks with mobile elements: a survey. *ACM Transactions on Sensor Networks*, 8(1), 2011.

[5] F. Koushanfar, M. Potkonjak, and A. Sangiovanni-Vincentell. Fault tolerance techniques for wireless ad hoc sensor networks. In *Proceedings of IEEE Sensors*, volume 2, pages 1491–1496, 2002.

[6] T. Melodia and I. F. Akyildiz. Research challenges for wireless multimedia sensor networks. In *Distributed Video Sensor Networks*. Springer, London, pages 233–246, 2011.

[7] S. Misra, T. Ojha, and A. Mondal. Game-theoretic topology control for opportunistic localization in sparse underwater sensor networks. *IEEE Transactions on Mobile Computing*, 14(5):990–1003, 2015.

[8] L. Mottola and G. P. Picco. Programming wireless sensor networks: fundamental concepts and state of the art. *ACM Computing Surveys*, 43(3), 2011.

[9] A. Munir, J. Antoon, and A. Gordon-Ross. Modeling and analysis of fault detection and fault tolerance in wireless sensor networks. *ACM Transactions on Embedded Computing Systems*, 14(1), 2015.

[10] M. S. Obaidat and S. Misra. Wireless underground sensor networks. In *Principles of Wireless Sensor Networks*. Cambridge University Press, Cambridge, UK, pages 348–370, 2014.

[11] N. A. Pantazis, S. A. Nikolidakis, and D. D. Vergados. Energy-efficient routing protocols in wireless sensor networks: a survey. *IEEE Communications Surveys and Tutorials*, 15(2):551–591, 2013.

[12] E. Sabbah and K. D. Kang. Security in wireless sensor networks. In *Guide to Wireless Sensor Network*. Springer, London, pages 491–512, 2009.

[13] A. Sangwan and R. P. Singh. Survey on coverage problems in wireless sensor networks. *Wireless Personal Communications*, 80(4):1475–1500, 2015.

[14] C. Schurgers, V. Tsiatsis, S. Ganeriwal, and M. Srivastava. Topology management for sensor networks: exploiting latency and density. In *Proceedings of ACM International Symposium on Mobile ad hoc Networking & Computing*, pages 135–145, Switzerland, 2002.

[15] A. R. Silva and M. C. Vuran. Communication with above-ground devices in wireless underground sensor networks: an empirical study. In *Proceedings of ICC*, pages 1–6, Cape Town, May 2010.

[16] F. Yang, K. C. Wang, and Y. Huang. Energy-neutral communication protocol for very low power microbial fuel cell based wireless sensor network. *IEEE Sensors Journal*, 15(4):2306–2315, 2015.

[17] J. Yick, B. Mukherjee, and D. Ghosal. Wireless sensor network survey. *Computer Networks*, 52:2292–2330, 2008.

[18] M. Younis, I. F. Senturk, K. Akkaya, S. Lee, and F. Senel. Topology management techniques for tolerating node failures in wireless sensor networks: a survey. *Computer Networks*, 58:254–283, 2014.

Chapter 2

Advances in Soft Computing

Debasis Samanta

Indian Institute of Technology Kharagpur

Gaurang Panchal

Indian Institute of Technology Kharagpur

2.1 Introduction

In the recent past, soft computing (SC) techniques were applied to complex and computationally intensive problems that have customarily proved intractable for conventional mathematical methods. Because of its proven strength in treating imprecise information and discovering novel solutions

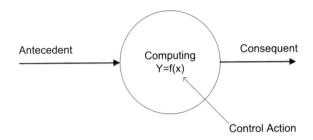

FIGURE 2.1: Concept of computing

to hard problems, SC is a topic of interest amongst researchers in various fields of science and engineering. Some popular SC paradigms include fuzzy computing [1], neural computing [6], and evolutionary computing [7]. In general, SC intends to generate computing mechanisms that demonstrate useful properties, that is, formulating inference with vague and/or ambiguous data, adapting to dynamic environments, learning from noisy and/or incomplete information, and reasoning with uncertainties.

Application of SC to real-world problems is, however, complicated, as it associates several key issues (i.e., learning models, knowledge representation, evaluation criteria, etc.) requiring crucial options among several approaches proposed in this domain. Further, results and lessons learnt from the application of SC to real-world problems are of major significance for methodological research with the intention of refining the existing methods or devising new techniques.

The overall aim of this chapter is to compile the recent advances in the field of SC. In particular, parallel and hybrid SC approaches are entertained and the chapter mainly focuses on the following.

1. Architectural design and development of fuzzy logic, neural networks, evolutionary computation

2. Comparative theoretical and empirical studies on SC techniques

3. Applications of SC techniques

2.1.1 Basic Concept of Soft Computing

In essence, the concept of soft computing implies mapping an input (x) to an output (y) given the mapping function $y = f(x)$ (also see Fig. 2.1). Here, input, output and mapping are called antecedent, consequent and control action, respectively. The characteristics of such a computing are

1. It should provide a precise solution.
2. The control action should be unambiguous and accurate.

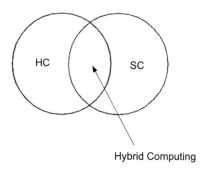

Hybrid Computing

FIGURE 2.2: Concept of hybrid computing

3. It is suitable for problems that are easy to model mathematically. That is, control actions can be mathematically modeled.

The characteristics as mentioned above were initially proposed by L. A. Zadeh in 1965. He called them "hard computing" instead of simple computing. Some examples of hard computing are: solving numerical problems (e.g., finding roots of a polynomial), searching and sorting of data, and computational geometry (e.g., finding area of any geometrical shape). One can easily argue that for computing such problems, precise results are guaranteed and control actions are unambiguous and can be modeled mathematically.

In contrast to hard computing, Zadeh coined the term "soft computing" for a system that does not satisfy any one or more characteristic(s) of hard computing. Some examples of soft computing are: medical diagnosis, person identification vis-a-vis computer vision, hand written character recognition, pattern recognition, anomaly detection, machine intelligence, weather forecasting, layout optimization and floor planning in VLSI design, network optimization, etc. The soft computing does not require any mathematical modeling of problem solving. It may not yield guaranteed precise solutions. Control actions are adaptive (i.e., can adjust to the change of dynamic environment). In general, control actions are conceived with some biological inspired methodologies such as genetics, evolution, and ant behaviors.

2.1.2 Hybrid Computing

It is a combination of the conventional hard computing and emerging soft computing (see Fig. 2.2). Hybrid systems that combine several soft computing techniques are needed to solve complex problems. Hybrid intelligent systems can have different architectures, which have an impact on the efficiency and accuracy of complex systems to achieve the ultimate goal of higher accuracy.

2.2   Fuzzy Logic

The human thought and decision-making process is linguistic. Our brains have evolved in such a way that if someone giving you directions says "the theater is a little further ahead to your left," you will automatically grasp and decipher the meaning of "a little further ahead." The person giving you directions doesn't need to say: "it's 31 meters left and then 29.4 meters from your current location."

Therefore, not only does an intelligent system need to be able to handle uncertainty, it should be able to interpret linguistic variables. Fuzzy logic gives a beautifully intuitive way to achieve this.

2.2.1   Basic Concept of Fuzzy Logic

The concept of fuzzy logic was proposed by L. A. Zadeh (1965). Using the fuzzy logic concept, truth value of a parameter can be mathematically represented by any real number between 0 and 1, instead of only 0 or 1. Therefore, this concept is widely used where the truth value of a parameter is neither completely true nor completely false. There are some other mathematical languages, for example,

1. Relational algebra (deals with manipulation of sets)
2. Boolean algebra (defines operations on Boolean variables)
3. Predicate logic (operations on well formed formulae (wff), also called predicate proposition)

In a similar manner fuzzy logic deals with fuzzy elements and fuzzy sets. We may note that the dictionary meaning of "fuzzy" is "unclear" or "uncertain." Its meaning can be better understood from its antonym, that is, "crisp."

We have already mentioned that fuzzy logic deals with fuzzy sets. To understand the concept of **fuzzy set** it is better to describe **crisp set**. Let $X =$ the entire population of India, $H =$ all Hindu population and $M =$ all Muslim population (Fig. 2.3). H $= \{h_1, h_2, h_3, ..., h_L\}$, M $= \{m_1, m_2, m_3, ..., m_N\}$. All

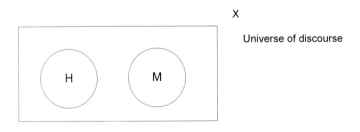

FIGURE 2.3: Universe of discourse

are sets of finite numbers of individuals. Here, H = {h | h follows all customs of Hindu religion}. Such a set is called crisp set. Now, let us discuss about fuzzy Set. X = all students in a class. S = all **Good students**, $S = \{(s, g) \mid s \in X\}$ and g(s) is a measurement of goodness of the student s.

Example: S = {(Rajat, 0.8), (Kabita, 0.7), (Salman, 0.1), (Ankit, 0.9)}.

Note: In case of a crisp set, the elements are with extreme values of degree of membership, namely either 1 or 0. On the other hand, the values of degree of memberships for a fuzzy element are within 0 and 1, both inclusive.

2.2.2 Fuzzy Logic Controller

Fuzzy logic can be applied to many applications, such as fuzzy reasoning, fuzzy clustering, and fuzzy programming. Amongst these applications, fuzzy reasoning, also called fuzzy logic controller (FLC), is an important application. Fuzzy logic controllers are special expert systems. In general, a FLC employs a knowledge base expressed in terms of a fuzzy inference rules and a fuzzy inference engine to solve a problem. We use FLC where an exact mathematical formulation of a problem is not possible or very difficult. These difficulties are due to non-linearities, time-varying nature of the process, large unpredictable environment disturbances, etc. A general fuzzy controller consists of four modules: a fuzzy rule base, a fuzzy inference engine, and the fuzzification and defuzzification modules (see Fig. 2.4). A general scheme of a fuzzy controller is shown in Fig. 2.4. As shown in the figure, a fuzzy controller operates by repeating a cycle of the following four steps:

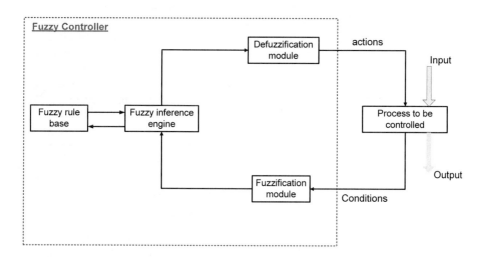

FIGURE 2.4: Fuzzy logic controller

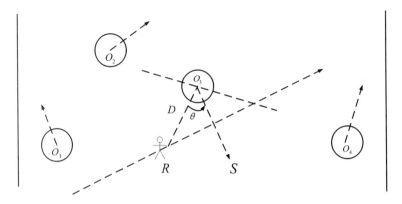

FIGURE 2.5: Navigation of robot

1. Measurements (inputs) are taken of all variables that represent relevant conditions of the controller process.

2. These measurements are converted into appropriate fuzzy sets to express measurement uncertainties. This step is called fuzzification.

3. The fuzzified measurements are then used by the inference engine to evaluate the control rules stored in the fuzzy rule base. The result of this evaluation is a fuzzy set (or several fuzzy sets) defined on the universe of possible actions.

4. This output fuzzy set is then converted into a single (crisp) value (or a vector of values). This is the final step called defuzzification. The defuzzified values represent actions to be taken by the fuzzy contoller.

Example of fuzzy logic controller: Consider the control of navigation of a robot (Fig. 2.5), in the presence of a number of moving objects. To make the problem simple, consider only four moving objects, each of equal size and moving with the same speed. A typical scenario is shown in Fig. 2.5. We consider two parameters: D, the distance from the robot to an object; θ the angle of motion of an object with respect to the most critical object will decide an output called deviation (δ). We assume the range of values of D is $[0.1, \dots 2.2]$ in meters and θ is $[-90, \dots, 0, \dots 90]$ in degrees. After identifying the relevant input and output variables of the controller and their range of values, the Mamdani approach to select some meaningful states called "linguistic states" for each variable and express them by appropriate fuzzy sets. For the current example, we consider the following linguistic states for the three parameters.

Linguistic states: Distance is represented by four linguistic states: VN = very near; NR = near; VF = very far; and FR = far. Angles for directions (θ) and deviation (δ) are represented by five linguistic states: LT = left; AL =

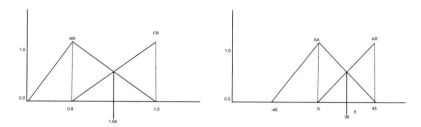

FIGURE 2.6: Fuzzification of inputs

ahead left; AR = ahead right; and RT = right. Three fuzzy sets for the three parameters are given in Fig. 2.6.

Fuzzy rule base: Once the fuzzy sets of all parameters are worked out, our next step in FLC design is to decide the fuzzy rule base of the FLC. The rule base for the FLC of mobile robot is shown in the form of a table below.

Fuzzy rule base for mobile robot: Note that this rule base defines 20 rules for all possible instances. These rules are simple and in the following form.

Rule 1: If (distance is VN) and (angle is LT), (deviation is AA).

⋮

⋮

Rule 13: If (distance is FR) and (angle is AA), (deviation is AR).

⋮

⋮

Rule 20: If (distance is VF) and (angle is RT), (deviation is AA).

2.2.2.1 Fuzzification

The next step is the fuzzification of inputs. Let us consider at any instant, the object O_3 is critical to the mobile robot and distance $D = 1.04m$ and angle $\theta = 30$ degrees (see Fig. 2.6). For this input, we are to decide the deviation δ of the robot as output. From the given fuzzy sets and input parameter values, we say that the distance $D = 1.04\ m$ may be called NR (near) or FR (far). Similarly, the input angle $\theta = 30$ degrees can be declared as either AA (ahead) or AR (ahead right).

Hence, we are to determine the membership values corresponding to these values as follows: $x = 1.04\ m$, $\mu_{NR}(x) = 0.6571$, $\mu_{FR}(x) = 0.3429$, $y = 30$ degrees, $\mu_{AA}(y) = 0.3333$, $\mu_{AR}(y) = 0.6667$.

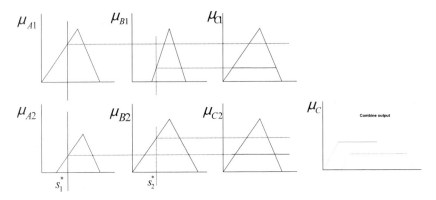

FIGURE 2.7: Fuzzy output

Rule strength computation: There are many rules in the rule base and not all rules may be applicable. For the given $x = 1.04$ and $\theta = 30$ degrees, only the following four rules out of 20 rules are possible.

R1: If (distance is NR) and (angle is AA), (deviation is RT).
R2: If (distance is NR) and (angle is AR), (deviation is AA).
R3: If (distance is FR) and (angle is AA), (deviation is AR).
R4: If (distance is FR) and (angle is AR), (deviation is AA).

The strength (also called α values) of the firable rules are calculated as follows.

$\alpha(R1) = min(\mu_{NR}(x), \mu_{AA}(y)) = min(0.6571, 0.3333) = 0.3333$
$\alpha(R2) = min(\mu_{NR}(x), \mu_{AR}(y)) = min(0.6571, 0.6667) = 0.6571$
$\alpha(R3) = min(\mu_{FR}(x), \mu_{AA}(y)) = min(0.3429, 0.3333) = 0.3333$
$\alpha(R4) = min(\mu_{FR}(x), \mu_{AR}(y)) = min(0.3429, 0.6667) = 0.3429$

Fuzzy output: The next step is to determine the fuzzified outputs corresponding to all fired rules. The working principle is first discussed and then we illustrate with a running example. Suppose only two fuzzy rules apply as below.

R1: IF (s_1 is A_1) AND (s_2 is B_1), (f is C_1).
R2: IF (s_1 is A_2) AND (s_2 is B_2), (f is C_2).

Suppose s_1^* and s_2^* are the inputs for fuzzy variables s_1 and s_2. μ_{A1}, μ_{A2}, μ_{B1}, μ_{B2}, μ_{C1} and μ_{C2} are the membership values for different fuzzy sets. This is graphically shown in Fig. 2.7.

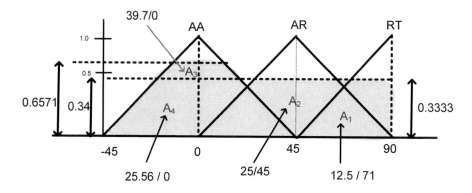

FIGURE 2.8: Combined fuzzified output for all four red rules

We take min of membership function values for each rule. Output membership function is obtained by aggregating the membership function of result of each rule. Fuzzy output is nothing but fuzzy OR of all outputs of rules.

2.2.2.2 Defuzzification

The fuzzy output needs to be defuzzified and its crisp value has to be determined for the output to make a decision. From the combined fuzzified output for all four fired rules (Fig. 2.8), we get the crisp value using the center of sum method as follows: $v = \frac{12.5 \times 71 + 25 \times 45 + 25.56 \times 0 + 25.56 \times 0}{12.5 + 39.79 + 25 + 25.56} = 19.59$.

Conclusion: The robot should deviate by 19.58089 degrees toward the right with respect to the line joining the move of direction to avoid collision with the obstacle O_3.

2.3 Artificial Neural Networks

An artificial neural network (ANN) is an interconnected group of nodes, akin to the vast network of neurons in a brain. In machine learning and cognitive science, ANNs are a family of statistical learning models inspired by biological neural networks (the central nervous systems of animals, in particular the brain) and are used to approximate functions that can depend on a large number of inputs and are generally unknown. Artificial neural networks are generally presented as systems of interconnected neurons that send messages to each other. The connections have numeric weights that can be tuned based on experience, making neural nets adaptive to inputs and capable of learning. This section introduces the underlying concepts and workings of ANNs.

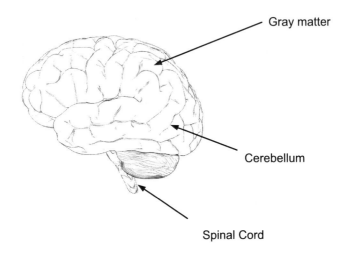

FIGURE 2.9: Human brain

2.3.1   Basic Concepts

A biological nervous system is an important component of most living things, in particular humans. A human brain (Fig. 2.9) is at the center of the human nervous system. In fact, any biological nervous system consists of a large number of interconnected processing units called neurons. Each neuron is approximately 10 μm long and they can operate in parallel. Typically, a human brain consists of approximately 10^{11} neurons communicating with each other with the help of electrical impulses. Figure 2.10 shows a biological neuron. There are different parts: dendrite, soma, axon, and synapse. **Dendrite**: bush of very thin fibre, **axon**: long cylindrical fibre, **soma**: also called a cell body, and acts like a cell nucleus, **synapse**: a junction where axon makes contact with the dendrites of neighboring dendrites.

There is a chemical in each neuron called a neurotransmitter. A signal (also called sense) is transmitted across neurons by this chemical. That is, all inputs from other neuron arrive to a neuron through dendrites. Unlike dendrite links, the action is electrically active and serves as an output channel. An action may produce an electrical impulse, which usually lasts for about a millisecond. Note that this pulse is generated by an incoming signal and a signal may not produce pulses in an axon unless it crosses a **threshold value**.

We may note that a neutron is a part of an interconnected network of a nervous system and provides services such as input signals, transportation of signals (at a very high speed), storage of information, perception, and automatic training and learning.

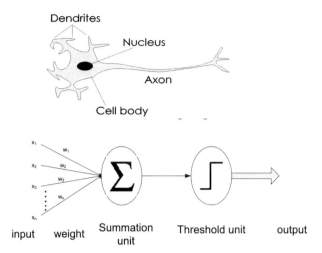

FIGURE 2.10: Artificial neuron and its input

In fact, the human brain is a highly complex structure viewed as a massive, highly interconnected network of simple processing elements called **neurons**. Artificial neural networks (ANNs) or simply neural networks (NNs), are simplified models of the biological nervous system, and can imitate the kind of computing performed by the human brain.

The behavior of a neuron can be captured by a simple model as shown in Fig. 2.10. Also, we can see the analogy between the biological neuron and artificial neuron. Truly, every component of the model (i.e., artificial neuron) bears a direct analogy to that of a biological neuron. It is this model which forms the basis of a neural network. In Fig. 2.10, $x_1, x_2, \cdots, x_n$ are the n inputs to the artificial neuron. The $w_1, w_2, \cdots, w_n$ are weights attached to the input links. Note that, a biological neuron receives all inputs through the dendrites, sums them and produces an output if the sum is greater than a threshold value. The input signals are passed on to the cell body through the synapse, which may accelerate or retard an arriving signal. This acceleration or retardation of the input signals is modeled by the **weights**. An effective synapse which transmits a stronger signal will have correspondingly larger weights while a weak synapse will have smaller weights. Thus, weights are multiplicative factors of the inputs to account for the strength of the synapse. Hence, the total input I received by the soma of the artificial neuron is

$$I = w_1 x_1 + w_2 x_2 + \cdots + w_n x_n = \sum_{i=1}^{n} w_i x_i \qquad (2.1)$$

To generate the final output y, the sum is passed to a filter ϕ called a transfer function, which releases the output. That is,

$$y = \phi(I) \qquad (2.2)$$

A common transfer function is the **thresholding function**. A sum (i.e., I) is compared with a threshold value θ. If the value of I is greater than θ, the output is 1 or 0 (like a simple linear filter). In other words,

$$y = \phi(\sum_{i=1}^{n} w_i x_i - \theta) \tag{2.3}$$

where $\phi(I) = \begin{cases} 1 & , \text{if } I > \theta \\ 0 & , \text{if } I \leq \theta \end{cases}$ Such a Φ is called step function (also known as a Heaviside function). Fig. 2.11 illustrates the thresholding function.

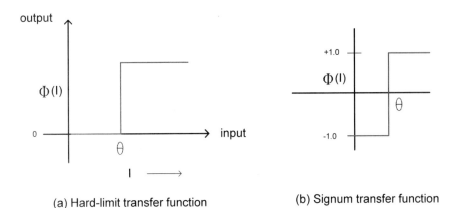

(a) Hard-limit transfer function   (b) Signum transfer function

FIGURE 2.11: Thresholding functions

Other transformation functions include:

Hard-limit transfer function: The transformation we have just discussed is called the hard-limit transfer function. It is generally used in perception neurons.

Linear transfer function: The output of the transfer function is made equal to its input (normalized) and lies in the range of -1.0 to $+1.0$. It is also known as the signum or quantizer function and it is defined as

$$\phi(I) = \begin{cases} +1 & , \text{if } I > \theta \\ -1 & , \text{if } I \leq \theta \end{cases} \tag{2.4}$$

Sigmoid transfer function: This function is continuous and varies gradually between the asymptotic values 0 and 1 (called log-sigmoid) or -1 and $+1$ (called tan-sigmoid) threshold functions as given by

$$\phi(I) = \frac{1}{1 + e^{-\alpha I}} [log - sigmoid] \tag{2.5}$$

$$\phi(I) = tanh(I) = \frac{e^{\alpha I} - e^{-\alpha I}}{e^{\alpha I} + e^{-\alpha I}} [tan - sigmoid] \qquad (2.6)$$

where α is the coefficient of transfer function. Such transfer function (Fig. 2.12) is used in back propagation neural networks (BPNNs).

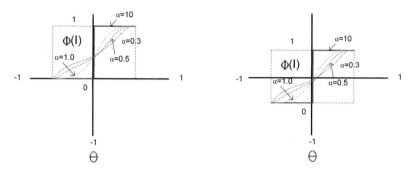

(a) Log-Sigmoid transfer function   (b) Tan-Sigmoid transfer function

FIGURE 2.12: Other thresholding functions

Advantages of Artificial Neural Networks:

- ANNs exhibit mapping capabilities, that is, they can map input patterns to their associated output pattern.

- ANNs learn by examples. Thus, an ANN architecture can be trained with a known example of a problem before it is tested for its inference capabilities on an unknown problem. In other words, ANNs can identify new objects without training.

- ANNs possess the capability to generalize. This is the power to apply a solution where an exact mathematical model is not possible.

- ANNs are robust and fault tolerant. They can, therefore, recall full patterns from incomplete, partial or noisy patterns.

- ANNs can process information in parallel, at high speed and in a distributed manner. Thus a massively parallel distributed processing system made of highly interconnected (artificial) neural computing elements having ability to learn and acquire knowledge is possible.

2.3.2   Neural Network Architecture

There are three fundamental classes of ANN architectures.

- Single layer feed forward architecture

- Multilayer feed forward architecture

- Recurrent network architecture

Before we discuss all these architecture, we first explain the mathematical details of a neuron at a single level. To do this, let us first consider the AND problem and its possible solution with a neural network. The simple Boolean AND operation with two input variables x_1 and x_2 is shown in the truth table (Table 2.1). Here, we have four input patterns: 00, 01, 10 and 11. For the first three patterns, output is 0 and for the last pattern output is 1.

TABLE 2.1: Boolean AND Operation

Inputs		Output
x_1	x_2	
0	0	0
0	1	0
1	0	0
1	1	1

Alternatively, the AND problem can be thought as a perception problem where we have to receive four different patterns as input and perceive the results as 0 or 1. A possible neuron specification to solve the AND problem is given in Fig. 2.13. In this solution when the input is 11, the weight sum exceeds the threshold ($\theta = 0.9$) leading to the output 1 or 0. Here, $y = \sum w_i x_i - \theta$

Inputs		Output (y)
x_1	x_2	
0	0	0
0	1	0
1	0	0
1	1	1

The AND Logic

A single neuron

FIGURE 2.13: Sample neuron

and $w_1 = 0.5$, $w_2 = 0.5$ and $\theta = 0.9$. The concept of the AND problem and its solution with a single neuron can be extended to multiple neurons. A layer of n neurons is shown in Fig. 2.13. This shows a single layer neural network. Note that the input layer and output layer which receive input signals and transmit output signals are called layers but they are actually boundaries of the architecture and hence truly not layers. The only layers in the architecture are the synaptic links carrying the weights that connect every input to the output neuron.

In a single layer neural network, the inputs $x_1, x_2, \cdots, x_m$ are connected to the layers of neurons through the weight matrix W. The weight matrix

$W_{m \times n}$ can be represented as follows.

$$
w = \begin{vmatrix}
w_{11} & w_{12} & w_{13} & \cdots & w_{1n} \\
w_{21} & w_{22} & w_{23} & \cdots & w_{2n} \\
\vdots & \vdots & \vdots & & \vdots \\
w_{m1} & w_{m2} & w_{m3} & \cdots & w_{mn}
\end{vmatrix}
\tag{2.7}
$$

The output of any k-th neuron can be determined as follows.

$$
O_k = f_k \left(\sum_{i=1}^{m} (w_{ik} x_i + \theta_k) \right)
\tag{2.8}
$$

where $k = 1, 2, 3, \cdots, n$ and θ_k denotes the threshold value of the k-th neuron. This network is feed forward in type or acyclic in nature and hence the name.

Multilayer feed forward neural networks: This network, as its name indicates, is made of multiple layers. Thus architectures of this class besides processing an input and an output layer also have one or more intermediary layers called hidden layers. The hidden layers aid in performing useful intermediary computation before directing the input to the output layer. A multilayer feed forward network with one input neurons (number of neuron at the first layer), $m_1, m_2, \cdots, m_p$ neurons at i-th hidden layer ($i = 1, 2, \cdots, p$) and n neurons at the last layers (output layer) is written as $l - m_1 - m_2 - \cdots - m_p - n$.

Multilayer feed forward neural networks: Figure 2.14 shows a multilayer feed forward neural network with a configuration of $l - m - n$. Here the input first layer contains l numbers of neurons, the only hidden layer contains m number of neurons and the (output) last layer contains n number of neurons. The inputs $x_1, x_2, \ldots x_p$ are fed to the first layer and the weight between the first and the hidden layer and those between hidden and the last layer are denoted as W^1, W^2, and W^3, respectively. Further, consider that f^1, f^2, and f^3 are the transfer functions of neurons lying on the first, hidden, and last layers respectively. Likewise, the threshold value of any i-th neuron in the j-th layer is denoted by θ_i^j. Moreover, the output of i-th, j-th, and k-th neurons in any l-th layer is represented by $O_i^l = f_i^l \left(\sum X_i W^l + \theta_i^l \right)$, where X_l is the input vector to the l-th layer.

Recurrent neural network architecture: The networks differ from feedback network architectures in the sense that there is at least one "feedback loop." Thus, in these networks, for example, there could exist one layer with feedback connections as shown in Fig. 2.14. There could also be neurons with self-feedback links, that is, the output of a neuron is fed back into itself as input. Depending on different types of feedback loops, several recurrent neural networks are known such as Hopfield network, Boltzmann machine network etc.

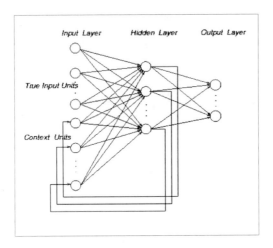

FIGURE 2.14: Recurrent neural network architecture

Why different types of neural network architecture? To give the answer to this question, let us first consider the case of a single neural network with two inputs as shown in Fig. 2.15. Note that $f = b_0 + w_1 x_1 + w_2 x_2$ denotes a straight line in the plane of x_1-x_2 (as shown at right). Now, depending on the values of values of w_1 and w_2, we have a set of points for different values of x_1 and x_2. We then say that these points are linearly separable, if the straight line f separates these points into two classes. Linearly separable and non-separable points are further illustrated in Figure 2.15.

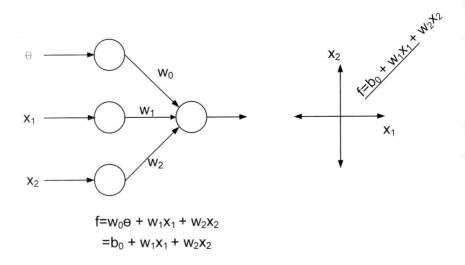

FIGURE 2.15: Linearly separable and non-separable points

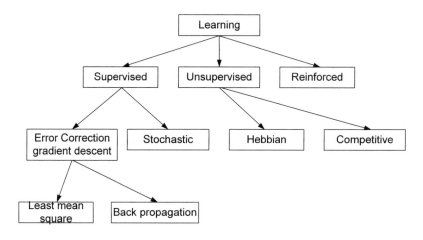

FIGURE 2.16: Classification of learning techniques

2.3.3   ANN Training and Learning

Learning is an important feature of human computational ability. Learning may be viewed as the change in behavior acquired due to practice or experience, and it lasts for a relatively long time. As learning occurs, the effective coupling between the neurons is modified. In case of artificial neural networks, it is a process of modifying the network by updating its weights, biases and other parameters, if any. During the learning, the parameters of the networks are optimized and as a result undergo curve fitting. It is then said that the network has passed through a learning phase.

2.3.3.1   Types of Learning

There are several learning techniques. A taxonomy of well known learning techniques is shown in Fig. 2.16. In the following, we discuss these learning techniques.

Supervised learning: Every input pattern that is used to train the network is associated with an output pattern. This is the "training set of data." Thus, in this form of learning, the input-output relationships of the training scenarios are available. The output of a network is compared with the corresponding target value and the error is determined. It is then fed back to the network for updating. This results in an improvement. This type of training is called learning with the help of a teacher.

Unsupervised learning: If the target output is not available, the error in prediction cannot be determined and in such a situation, the system learns on its own by discovering and adapting to structural features in the input patterns. This type of training is called learning without a teacher.

Reinforced learning: Although a teacher is available, it does not reveal the expected answer, but does so only if the computed output is correct or incorrect. A reward is given for a correct answer computed and a penalty results from a wrong answer. This information helps the network in its learning process.

Note: Supervised and unsupervised learnings are the most popular forms. Unsupervised learning is very common in biological systems. It is also important for artificial neural networks because training data are not always available for the intended application of the neural network. Other useful learning techniques are described below.

Hebbian learning: This learning is based on correlative weight adjustment. This is, in fact, the learning technique inspired by biology. Here, the input-output pattern pairs (x_i, y_i) are associated with the weight matrix W. W is also known as the correlation matrix. This matrix is computed as follows.

$$W = \sum_{i=1}^{n} X_i Y_i^T \qquad (2.9)$$

where Y_i^T is the transpose of the associated vector y_i

Gradient descent learning: This learning technique is based on the minimization of error E defined in terms of weights and the activation function of the network. Also, it is required that the activation function employed by the network is differentiable, as the weight update is dependent on the gradient of the error E. Thus, if ΔW_{ij} denoted the weight update of the link connecting the i-th and j-th neuron of the two neighboring layers then

$$\Delta W_{ij} = \eta \frac{\partial E}{\partial W_{ij}} \qquad (2.10)$$

where η is the learning rate parameter and $\frac{\partial E}{\partial W_{ij}}$ is the error gradient with reference to the weight W_{ij}. The least mean square and back propagation are two variations of this learning technique.

Competitive learning: The neurons which respond strongly to input stimuli have their weights updated. When an input pattern is presented, all neurons in the layer compete and the winning neuron undergoes weight adjustment. This is why it is called a winner-takes-all strategy.

Stochastic learning: Weights are adjusted in a probabilistic fashion. Simulated annealing is an example of such learning (proposed by Boltzmann and Cauchy). In the following, we discuss a generalized approach to train different types of neural network architectures.

2.3.3.2 Single Layer Feed Forward NN Training

We know that several neurons are arranged in one layer with inputs and weights connect to every neuron. Learning in such a network occurs by adjusting the weights associated with the inputs so that the network can classify the input patterns. A single neuron in such a neural network is called a perceptron. The algorithm to train a perceptron is stated below. Consider a perceptron with $(n + 1)$ inputs $x_0, x_1, x_2, \cdots, x_n$ where $x_0 = 1$ is the bias input. Let f denote the transfer function of the neuron. Suppose, $\bar{X}$ and $\bar{Y}$ denote the input and output vectors as a training data set. $\bar{W}$ denotes the weights vector.

With this input-output relationship pattern and configuration of a perceptron, the algorithm to train the perceptron is stated below.

1. Initialize $\bar{W} = w_0, w_1, \cdots, w_n$ to some random weights.

2. For each input pattern $x \in X$

   - Compute $I = \sum_{i=0}^{n} w_i x_i$.
   - Compute observed output y

   $$y = f(I) = \begin{cases} 1 & , \text{ if } I > 0 \\ 1 & , \text{ if } I \leq 0 \end{cases}$$

   $$\bar{Y}' = \bar{Y}' + y$$

3. If the desired output $\bar{y}$ matches the observed output $\bar{y}'$, output $\bar{W}$ and exit.

4. Otherwise, update the weight matrix $\bar{W}$ as follows:

   - For each output $y \in \bar{Y}'$:
   - If the observed out y is 1 instead of 0, then $w_i = w_i - \alpha - x_i$, $(i = 1, 2, \cdots n)$.
   - If the observed out y is 0 instead of 1, then $w_i = w_i + \alpha - x_i$, $(i = 1, 2, \cdots n)$.

5. Go to step 2.

In the above algorithm, α is the learning parameter and is a constant decided by some empirical studies.

Note: The training perceptron algorithm is based on supervisory learning techniques. An alternate term for perceptron is adaptive linear network element (ADALINE). If ten neutrons in a single layer forward feed neural network are to be trained, the algorithm must be iterated for each perceptron in the network.

As for a single layer feed forward neural network, supervisory training methodology is followed to train a multilayer feed forward neural network. Before we try to understand the training of such a neural network, we redefine some terms.

For simplicity, we assume that all neurons in a particular layer follow the same transfer function and different layers follow their respective transfer functions as shown in the above figure. Let us consider a specific neuron in each layer, say i-th, j-th and k-th neurons in the input, hidden and output layer, respectively. Also, let us denote the weight between the i-th neuron $(i = 1, 2, \cdots, l)$ in the input layer to j-th neuron $(j = 1, 2, \cdots, m)$ in the hidden layer by v_{ij}. The weight matrix between the input to hidden layer, say V, is denoted as follows.

$$V = \begin{bmatrix} v_{11} & v_{12} & \cdots & v_{1j} & \cdots & v_{1m} \\ v_{21} & v_{22} & \cdots & v_{2j} & \cdots & v_{2m} \\ \vdots & \vdots & \vdots & \vdots & \vdots & \vdots \\ v_{i1} & v_{i2} & \cdots & v_{ij} & \cdots & v_{im} \\ v_{l1} & v_{l2} & \cdots & v_{lj} & \cdots & v_{lm} \end{bmatrix} \tag{2.11}$$

Similarly, W_{ij} represents the connecting weights between the j-th neuron $(j = 1, 2, \cdots, m)$ in the hidden layer and k-th neuron $(k = 1, 2, \cdots n)$ in the output layer as follows.

$$W = \begin{bmatrix} w_{11} & w_{12} & \cdots & w_{1k} & \cdots & w_{1n} \\ w_{21} & w_{22} & \cdots & w_{2k} & \cdots & w_{2n} \\ \vdots & \vdots & \vdots & \vdots & \vdots & \vdots \\ w_{j1} & w_{j2} & \cdots & w_{jk} & \cdots & w_{jn} \\ w_{m1} & w_{m2} & \cdots & w_{mk} & \cdots & w_{ln} \end{bmatrix} \tag{2.12}$$

Input layer computation: Let $< T_0, T_I >$ be the training set of size $|T|$. Let us consider an input training data at any instant to be $I^I = [I_1^1, I_2^1, \cdots, I_i^1, I_l^1]$ where $I^I \in T_I$. Consider the outputs of the neurons lying on the input layer are the same as the corresponding inputs. That is,

$$O^I = I^I, [l \times 1] = [l \times 1] \tag{2.13}$$

Output of the input layer: The input of the j-th neuron in the hidden layer can be calculated as follows.

$$I_j^H = v_{1j}o_1^I + v_{2j}o_2^I +, \cdots, +v_{ij}o_j^I + \cdots + v_{ij}o_l^I \tag{2.14}$$

where $j = 1, 2, \cdots m$.

Calculation of input of each node in the hidden layer:
In the matrix representation form, we can write

$$I^H = V^T \cdot o^I, [m \times 1][m \times l][l \times 1] \tag{2.15}$$

Hidden layer computation: Let us consider any j-th neuron in the hidden layer. Since the outputs of the input layer's neurons are the inputs to the j-th neuron and the j-th neurons follow the log-sigmoid transfer function, we have

$$O_j^H = \frac{1}{1 + e^{-\alpha_H \cdot I_j^H}} \tag{2.16}$$

where $j = 1, 2, \cdots, m$ and α_H is the constant co-efficient of the transfer function.

Note that all outputs of the nodes in the hidden layer can be expressed as a one-dimensional column matrix.

$$O^H = \begin{bmatrix} \cdots \\ \cdots \\ \vdots \\ \frac{1}{1 + e^{-\alpha_H \cdot I_j^H}} \\ \vdots \\ \cdots \\ \cdots \end{bmatrix}_{m \times 1} \tag{2.17}$$

Output layer computation: Let us calculate the input to any k-th node in the output layer. Since outputs of all nodes in the hidden layer go to the k-th layer with weights $w_{1k}, w_{2k}, \cdots, w_{mk}$, we have

$$I_k^o = w_{1k} \cdot o_1^H + w_{2k} \cdot o_2^H + \cdots + w_{mk} \cdot o_m^H \tag{2.18}$$

where $k = 1, 2, \cdots, n$. In the matrix representation, we have

$$I^o = W^T \cdot O^H, [m \times 1][n \times m][m \times 1] \tag{2.19}$$

Now we estimate the output of the k-th neuron in the output layer. We consider the tan-sigmoid transfer function.

$$O_k = \frac{e^{\alpha_o \cdot I_k^o} - e^{-\alpha_o \cdot I_k^o}}{e^{\alpha_o \cdot I_k^o} + e^{-\alpha_o \cdot I_k^o}} \tag{2.20}$$

for $k = 1, 2, \cdots, n$. Hence, the output of the layer's neurons can be represented as

$$O = \begin{bmatrix} \cdots \\ \cdots \\ \vdots \\ \frac{e^{\alpha_o \cdot I_k^o} - e^{-\alpha_o \cdot I_k^o}}{e^{\alpha_o \cdot I_k^o} + e^{-\alpha_o \cdot I_k^o}} \\ \vdots \\ \cdots \\ \cdots \end{bmatrix}_{n \times 1} \tag{2.21}$$

The above discussion explains how to calculate values of different parameters in $l - m - n$ multiple layer feed forward neural networks. Next, we will discuss how to train such a neural network. We consider the most popular algorithm called back-propagation which is a supervised learning. The principle of the **back-propagation algorithm** is based on the error-correction with the **steepest-descent method**. We first discuss the method of steepest descent followed by its use in the training algorithm.

Method of steepest descent: Supervised learning is, in fact, error-based learning. In other words, with reference to an external (teacher) signal (i.e., target output) it calculates errors by comparing the target output and computed output. Based on the error signal, the neural network should modify its configuration, which includes synaptic connections, that is, its weight matrices. It should try to reach a state which yields minimum error. In other words, its searches for suitable values of parameters minimizing error, given a training set. Note that this turns out as an optimization problem.

For simplicity, let us consider the connecting weights are the only design parameters. Suppose, V and W are the weight parameters to hidden and output layers, respectively. Thus, given a training set of size N, the error surface, E can be represented as

$$E = \sum_{i=1}^{N} E^i (V, W, I_i) \qquad (2.22)$$

where I_i is the i-th input pattern in the training set. Now, we will discuss the steepest descent method of computing error, given changes in V and W matrices.

Suppose, A and B are two points on the error surface. The vector $\vec{AB}$ can be written as

$$\vec{AB} = (V_{i+1} - V_i) \cdot \bar{x} + (W_{i+1} - W_i) \cdot \bar{y} = \Delta V \cdot \bar{x} + \Delta W \cdot \bar{y} \qquad (2.23)$$

The gradient of $\vec{AB}$ can be obtained as

$$e_{\vec{AB}} = \frac{\partial E}{\partial V} \cdot \bar{x} + \frac{\partial E}{\partial W} \cdot \bar{y} \qquad (2.24)$$

Hence, the unit vector in the direction of gradient is

$$\bar{e}_{\vec{AB}} = \frac{1}{|e_{\vec{AB}}|} \left[\frac{\partial E}{\partial V} \cdot \bar{x} + \frac{\partial E}{\partial W} \cdot \bar{y} \right] \qquad (2.25)$$

With this, we can alternatively represent the distance vector AB as

$$\vec{AB} = \eta \left[\frac{\partial E}{\partial V} \cdot \bar{x} + \frac{\partial E}{\partial W} \cdot \bar{y} \right] \qquad (2.26)$$

where $\eta = \frac{k}{|e_{\overrightarrow{AB}}|}$ and k is a constant. So, comparing both, we have

$$\Delta V = \eta \frac{\partial E}{\partial V}, \Delta W = \eta \frac{\partial E}{\partial W} \tag{2.27}$$

This is also called the delta rule and η is called the learning rate.

Calculation of error in a NN: Let us consider any k-th neuron at the output layer. For an input pattern $I_i \in T_I$ (input in training) the target output T_{Ok} of the k-th neuron is T_{Ok}. Then, the error E_k of the k-th neuron is defined corresponding to the input I_i as

$$E_k = \frac{1}{2}(T_{Ok} - O_{Ok})^2 \tag{2.28}$$

where O_{Ok} denotes the observed output of the k-th neuron.

For a training session with $I_i \in T_I$, the error in prediction considering all output neurons can be given as

$$E = \sum_{k=1}^{n} e_k = \frac{1}{2} \sum_{k=1}^{n} (T_{Ok} - O_{Ok}) \tag{2.29}$$

where n denotes the number of neurons at the output layer. The total error in prediction for all output neurons can be determined considering all training session $< T_I, T_O >$ as $E = \sum_{\forall I_i \in T_I} e = \frac{1}{2} \sum_{\forall t \in <T_I, T_O>} \sum_{k=1}^{n} (T_{Ok} - O_{Ok})^2$.

Supervised learning: back-propagation algorithm

The back-propagation algorithm can be followed to train a neural network to set its topology, connecting weights, bias values and many other parameters. In this present discussion, we will only consider updating weights. Thus, we can write the error E corresponding to a particular training scenario T as a function of the variable V and W. That is, $E = f(V, W, T)$. In back-propagation algorithm, this error E is to be minimized using the gradient descent method. We know that according to the gradient descent method, the changes in weight value can be given as

$$\Delta V = -\eta \frac{\partial E}{\partial V} \tag{2.30}$$

and

$$\Delta W = -\eta \frac{\partial E}{\partial E} \tag{2.31}$$

Note that $-Ve$ sign is used to signify the fact that if $\frac{\partial E}{\partial V}$ (or $\frac{\partial E}{\partial W}$) > 0, then we have to decrease V and vice versa. Let v_{ij} (and w_{jk}) denote the weights connecting the i-th neuron (at the input layer) to the j-th neuron(at the hidden layer) and connecting the j-th neuron (at the hidden layer) to the k-th neuron (at the output layer). Also, let E_k denotes the error at the k-th neuron with observed output as $O_{O_k^o}$ and target output $T_{O_k^o}$ as per a sample input $I \in T_I$.

It follows logically that

$$E_k = \frac{1}{2}(T_{O_k^o} - O_{O_k^o})^2 \tag{2.32}$$

and the weight components should be updated as follows,

$$\bar{W}_{jk} = W_{jk} + \Delta W_{jk} \tag{2.33}$$

where $\Delta W_{jk} = \eta \frac{\partial E_k}{\partial W_{jk}}$
and

$$\bar{V}_{ij} = V_{ij} + \Delta V_{ij} \tag{2.34}$$

where $\Delta V_{ij} = \eta \frac{\partial E_k}{\partial V_{ij}}$. Here, v_{ij} and w_{ij} denote the previous weights and $\bar{v}_{ij}$ and $\bar{w}_{ij}$ denote the updated weights. Now we will learn the calculation $\bar{w}_{ij}$ and $\bar{v}_{ij}$, which is stated as below.

Calculation of $\bar{W}_{jk}$: We can calculate $\frac{\partial E_k}{\partial W_{jk}}$ using the chain rule of differentiation as given below.

$$\frac{\partial E_k}{\partial w_{jk}} = \frac{\partial E_k}{\partial O_{ok}^o} \cdot \frac{\partial O_{ok}^o}{\partial I_k^o} \cdot \frac{\partial I_k^o}{\partial w_{jk}} \tag{2.35}$$

Now,

$$E_k = \frac{1}{2}(T_{O_k^o} - O_{O_k^o})^2 \tag{2.36}$$

$$O_k^o = \frac{e^{oI_k^o} - e^{-oI_k^o}}{e^{oI_k^o} + e^{-oI_k^o}} \tag{2.37}$$

$$I_k^o = w_{1k} \cdot O_1^H + w_{2k} \cdot O_2^H + \cdots + w_{jk} \cdot O_j^H + \cdots w_{mk} \cdot O_m^H \tag{2.38}$$

Thus,

$$\frac{\partial E_k}{\partial O_{ok}^o} = -(T_{O_k^o} - O_{O_k^o}) \tag{2.39}$$

$$\frac{\partial O_k^o}{\partial I_k^o} = \theta_o(1 + O_{ok}^o)(1 - O_{ok}^o) \tag{2.40}$$

and

$$\frac{\partial I_k^o}{\partial w_{ij}} = O_j^H \tag{2.41}$$

Substituting the values of $\frac{\partial E_k}{\partial O_{ok}^o}$, $\frac{\partial O_{ok}^o}{\partial I_k^o}$ and $\frac{\partial I_k^o}{\partial w_{jk}}$ we have

$$\frac{\partial E_k}{\partial w_{jk}} = -(T_{O_k^o} - O_{O_k^o}) \cdot \theta_o(1 + O_{ok}^o)(1 - O_{ok}^o) \cdot O_j^H \tag{2.42}$$

Again, substituting the value of $\frac{\partial E_k}{\partial w_{jk}}$. We have

$$\Delta w_{jk} = \eta \cdot \theta_o(T_{O_k^o} - O_{O_k^o}) \cdot (1 + O_{ok}^o)(1 - O_{ok}^o) \cdot O_j^H \tag{2.43}$$

Therefore, the updated value of w_{jk} can be obtained as follows.

$$\bar{w_{jk}} = w_{jk} + \Delta w_{jk} = \eta \cdot \theta_o (T_{O_k^o} - O_{o_k^o}) \cdot (1 + O_{o_k^o})(1 - O_{o_k^o}) \cdot O_j^H + w_{jk} \quad (2.44)$$

Calculation of v_{ij}: Like, $\frac{\partial E_k}{\partial w_{jk}}$, we can calculate $\frac{\partial E_k}{\partial v_{ij}}$ using the chain rule of differentiation as follows,

$$\frac{\partial E_k}{\partial v_{ij}} = \frac{\partial E_k}{\partial O_{o_k^o}} \cdot \frac{\partial O_{o_k^o}}{\partial I_k^o} \cdot \frac{\partial I_k^o}{\partial O_j^H} \cdot \frac{\partial O_j^H}{\partial I_j^H} \cdot \frac{\partial I_j^H}{\partial v_{ij}} \quad (2.45)$$

Now,

$$E_k = \frac{1}{2}(T_{O_k^o} - O_{O_k^o})^2 \quad (2.46)$$

$$O_k^o = \frac{e^{\theta_o I_k^o} - e^{-\theta_o I_k^o}}{e^{\theta_o I_k^o} + e^{-\theta_o I_k^o}} \quad (2.47)$$

$$I_k^o = w_{1k} \cdot O_1^H + w_{2k} \cdot O_2^H + \cdots + w_{jk} \cdot O_j^H + \cdots w_{mk} \cdot O_m^H \quad (2.48)$$

$$O_j^H = \frac{1}{1 + e^{-\theta_H I_j^H}} \quad (2.49)$$

$$I_j^H = v_{ij} \cdot O_1^H + v_{2j} \cdot O_2^H + \cdots + v_{ij} \cdot O_j^I + \cdots v_{ij} \cdot O_l^I \quad (2.50)$$

Thus

$$\frac{\partial E_k}{\partial O_{o_k^o}} = -(T_{O_k^o} - O_{O_k^o}) \quad (2.51)$$

$$\frac{\partial O_k^o}{\partial I_k^o} = \theta_o (1 + O_{o_k^o})(1 - O_{o_k^o}) \quad (2.52)$$

$$\frac{\partial I_k^o}{\partial O_j^H} = w_{ik} \quad (2.53)$$

$$\frac{\partial O_j^H}{\partial I_j^H} = \theta_H \cdot (1 - \theta_j^H) \cdot \theta_j^H \quad (2.54)$$

from the above equation

$$\frac{\partial E_k}{\partial v_{ij}} = -\theta_o \cdot \theta_H (T_{O_k^o} - O_{O_k^o}) \cdot (1 - O_{O_k}^{2^o}) \cdot O_j^H \cdot I_j^H \cdot w_{jk} \quad (2.55)$$

substituting the value of $\frac{\partial E_k}{\partial v_{ij}}$, we have

$$\Delta v_{ij} = -\eta \cdot \theta_o \cdot \theta_H (T_{O_k^o} - O_{O_k^o}) \cdot (1 - O_{O_k}^{2^o}) \cdot O_j^H \cdot I_j^H \cdot w_{jk} \quad (2.56)$$

Therefore, the updated value of v_{ij} can be obtained as follows.

$$\bar{v_{ij}} = v_{ij} + \eta \cdot \theta_o \cdot \theta_H (T_{O_k^o} - O_{O_k^o}) \cdot (1 - O_{O_k}^{2^o}) \cdot O_j^H \cdot I_j^H \cdot w_{jk} \quad (2.57)$$

Writing in matrix form the calculation of $\bar{V}$ and $\bar{W}$

we have

$$\Delta w_{jk} = \eta \left| \theta_o \cdot (T_{O_k}^o - O_{O_k}^o) \cdot (1 - O_{O_k}^{2\,o}) \right| \cdot O_j^H \tag{2.58}$$

as the update for the k-th neuron receiving the signal from j-th neuron at the hidden layer.

$$\Delta v_{ij} = \eta \cdot \theta_o \cdot \theta_H (T_{O_k}^o - O_{O_k}^o) \cdot (1 - O_{O_k}^{2\,o}) \cdot O_j^H \cdot I_j^H \cdot w_{jk} \tag{2.59}$$

is the update for the j-th neuron at the hidden layer for the i-th input at the i-th neuron at input level.

Hence,

$$[\Delta W]_{m \times n} = \eta \cdot [O^H]_{m \times 1} \cdot [N]_{1 \times n} \tag{2.60}$$

where

$$[N]_{1 \times n} = \left\{ \theta_o (T_{O_k}^o - O_{O_k}^o) \cdot (1 - O_{O_k}^{2\,o}) \right\} \tag{2.61}$$

where $k = 1, 2, \cdots n$ Thus, the updated weight matrix for a sample input can be written as

$$[\bar{W}]_{m \times n} = [W]_{m \times n} + [\Delta W]_{m \times n} \tag{2.62}$$

Similarly, for $[\bar{V}]$ matrix, we can write

$$\Delta v_{ij} = \eta \cdot \left| \theta_o (T_{O_k}^o - O_{O_k}^o) \cdot w_{jk} \right| \cdot \left| \theta_H (1 - O_{O_k}^{2\,o}) \cdot O_j^H \right| \cdot \left| I_j^H \right| \tag{2.63}$$

$$= \eta \cdot w_j \cdot \theta^H \cdot (1 - O_j^H) \cdot O_j^H \tag{2.64}$$

Thus,

$$\Delta V = [I^I]_{l \times 1} \times [M^T]_{1 \times m} \tag{2.65}$$

or

$$[\bar{V}]_{l \times m} = [V]_{l \times m} + [I^I]_{l \times 1} \times [M^T]_{1 \times m} \tag{2.66}$$

This calculation is done for $t \in < T_O, T_I >$. We can apply it in incremental mode (i.e., one sample after another) and after each training data, we update the network V and W matrix.

Batch mode of training: A batch mode of training is generally implemented through the minimization of mean square error (MSE) in error calculation. The MSE for the k-th neuron at the output level is given by

$$\bar{E} = \frac{1}{2} \cdot \frac{1}{|T|} \sum_{t=1}^{|T|} \left(T^t{}_{O_k}^o - O^t{}_{O_k}^o \right)^2 \tag{2.67}$$

where $|T|$ denotes the total number of training scenario and t denotes a training scenario, i.e., $t \in < T_O, T_I >$. In this case, Δw_{jk} and Δv_{ij} can be calculated as follows

$$\Delta w_{jk} = \frac{1}{|T|} \sum_{\forall t \in T} \frac{\partial \bar{E}}{\partial W} \tag{2.68}$$

and

$$\Delta v_{ij} = \frac{1}{|T|} \sum_{\forall t \in T} \frac{\partial \bar{E}}{\partial V} \qquad (2.69)$$

Once Δw_{jk} and Δv_{ij} are calculated, we will be able to obtain w_{jk}^- and v_{ij}^-.

2.4 Evolutionary Algorithms

Evolutionary algorithms form a subset of evolutionary computation in that they generally only involve techniques implementing mechanisms inspired by biological evolution such as reproduction, mutation, recombination, natural selection, and survival of the fittest. Candidate solutions to the optimization problem play the role of individuals in a population, and the cost function determines the environment within which the solutions "live." Evolution of the population then takes place after the repeated application of the above operators.

In this process, there are two main forces that form the basis of evolutionary systems: recombination and mutation create the necessary diversity and thereby facilitate novelty, while selection acts as a force increasing quality.

2.4.1 Traditional Approaches to Optimization Problems

The traditional approaches to optimization problems are shown in Fig. 2.17.

Limitations of traditional optimization approach:

- It may trap into local optima.

- It is computationally expensive.

- For a discontinuous objective function, the method may fail

$$y = f(x), \forall x \qquad (2.70)$$

$y = f_1(x), \forall x \in X_1$ and $y = f_2(x), \forall x \in X_2$

$$y = \begin{cases} f_1(x) & , \forall x \in X_1 \\ f_2(x) & , \forall x \in X_2 \end{cases}$$

- The method may not be suitable for parallel computing.

- Discrete (integer) variable are difficult to handle.

- Methods may not be adaptive.

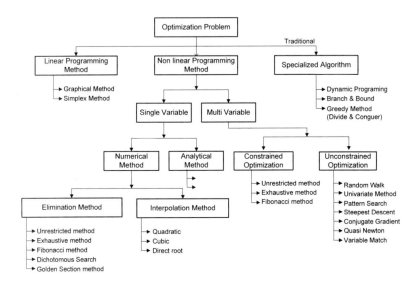

FIGURE 2.17: Traditional approaches to solve optimization problem

2.4.1.1 Evolutionary Computing

Evolutionary computing is the collective name for a range of problem-solving techniques based on principles of biological evolution, such as natural selection and genetic inheritance. These techniques are widely applied to a variety of problems. These techniques are shown in Fig. 2.18.

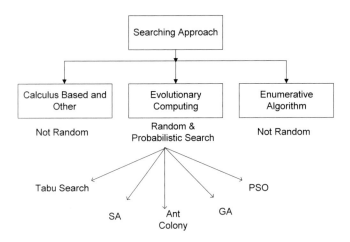

FIGURE 2.18: Types of searching algorithm

2.4.1.2 Genetic Algorithm (GA)

Genetic entities are body, cell chromosome (e.g., mosquito, human, frog, goldfish), gene, offspring (new chromosomes), crossover, and mutation (Fig. 2.18). The GA framework (see Fig. 2.19) involves evolution by selecting the best solution. Evolution involves the four primary premises listed below.

- Information propagation: An offspring has many characteristics of its parents (predecessors).

- Population with diversity leads to: variation in characteristics in the next generation.

- Survival for existence: only a small percentage of the offspring produced survival to adulthood.

- Survival for best: offspring survival depends on their inherited characteristics.

Definition: GA is a population-based probabilistic search and optimization technique based on the mechanisms of natural genetics and natural selection. The probabilistic search is iterative, the working cycle is with or without convergence and the solution is not guaranteed to be optimal. For an optimization problem, we require objective function, constraints input parameter, fitness evaluation, encoding, and decoding.

A population is a collection of individuals that represent possible candidate solutions, for example, a set of fruits of a certain weight. Inputs are fruits and weights. Examples of GA operators are encoding, convergence, mating pool, fitness evaluation, crossover, mutation, and inversion. The encoding schemes include binary encoding, real value encoding, order encoding, and tree encoding.

There are some hidden parameters:

Initial population size N
Size of mating pool p% of N
Convergence threshold δ (delta)
Mutation μ
Inversion η
Crossover R (set of crossover)

2.4.2 Multiobjective Optimization

Let us consider without loss of generality a multiobjective optimization problem with n decision variable and k objective function

$$\text{Minimize } y = f(x) = [y_1 \in f_1(x), y_2 \in f_2(x), \cdots, y_k \in f_k(x)]$$

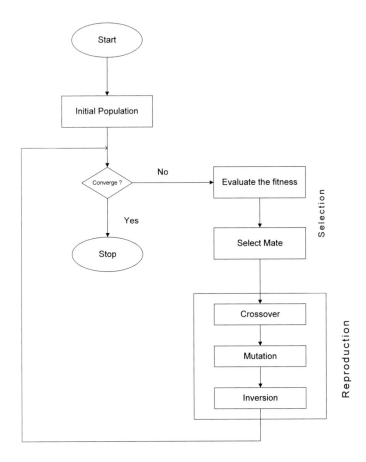

FIGURE 2.19: Framework of simple genetic algorithm

where

$$x = [x_1, x_2, \cdots, x_n] \in X, \ y = [y_1, y_2, \cdots, y_n] \in Y$$

Here, x is called a decision vector, y is called an objective vector, X is called a parameter space, and Y is called an objective space.

About the solution statement: We wish to determine $\bar{x} \in X$ (called feasible region in X) and any point $\hat{x} \in \bar{x}$ (which satisfy all the constraints in each f_i') is called a feasible solution. Also, we wish to determine from among the set $\hat{x}$, a particular solution $\hat{x}^*$ that yields the optimum values of the objective function. In other words, $\forall \hat{x} \in \bar{x}$ and $\exists \hat{x}^* \in \hat{x} \mid f_i(\hat{x}^*) \leq f_i(\hat{x})$, where $\forall_i \in [1, 2, \cdots, k]$.

Multiobjective optimization problem(MOOP): An optimization problem can be formally stated as follows.

Objectives (F)

$$\text{Minimize (maximize) } f_i(x_1, x_2, \cdots, x_n), \ i = 1, 2, \cdots, m$$

Constraints (S)

$$\text{Subject to } g_j(x_1, x_2, \cdots, x_n), \ ROP_j \ C_j, \ j = 1, 2, \cdots, l$$

Design variables (V)

$$\text{Subject to } x_k \ ROP_k \ d_k, \ k = 1, 2, \cdots, n$$

Note: For a MOOP, $m \geq 2$. Objective functions are of both minimization and maximization.

Multiobjective optimization problem (MOOP): why solving MOOP is an issue.

- In a single-objective optimization problem, task is to find the one solution which optimizes the sole objective function.

- In contrast, in MOOP

  - Cardinality of the optimal set is usually more than one.
  - There are $m \geq 2$ goals of optimization instead of one.
  - It possess $m \geq 2$ different search space.

- Optimizing each objective individually does not necessarily give the optimum solution.

  - Possible only if objective functions are independent on their solution space.

- Majority of real-world MOOPs have a set of trade-off optimal solutions.

  - In a particular search point one may be best whereas another may be worst.

- Also, some MOOPs have conflicting objectives.

  - Optimizing an objective means compromising other(s) and vice versa.

Search Space

Pareto optimal solution: A decision vector $\hat{x}^* \in \bar{x}$ is called a Pareto optimal if for every $\hat{x} \in \bar{x}$

$$f_i(\hat{x}^*) \leq f_i(\hat{x}) \text{ , for all i=1,2...k}$$

and

$$f_i(\bar{x}^*) < f_i(\hat{x}) \text{ , there exist at least one } f_i \in f$$

We defined Pareto optimum for the objective function to be minimized. The definition implies that $\hat{x}^*$ is a Pareto optimal solution and there exists no feasible vector $\hat{x}$ that decreases some objective function without causing a simultaneous increase in at least one other objective function. Pareto optimum may not necessarily give a single solution but rather a set of solutions called a non-dominated solution. A set of non-dominated solutions is called Pareto optimal set. All solutions in the set lie on the boundary (or in the loci of tangent points) of the feasible region. The region of points on which all non-dominated solutions lie is called Pareto front.

Evolutionary algorithm: GA-based multi-objective optimization techniques are as follows:

1. Naive approach

   - Weighted sum approach (single objective evolutionary algorithm (SOEA))
   - $\in$ constraint method
   - Goal attainment method

2. Non-aggregating approach

   - Vector evaluated genetic algorithm (VEGA)
   - Game theoretic approach
   - Lexicographic ordering
   - Weighted min-max approach

3. Pareto-based approach

   - Multiobjective genetic algorithm (MOGA)
   - Non-dominated sorting genetic algorithm (NSGA)
   - Niched Pareto genetic algorithm (NPGA)

Pareto-based approach: multiobjective optimization genetic algorithm (MOGA): There are two steps:

1. Rank assignment

   (a) All non-dominated solutions are assigned rank 1.

   (b) An ith iteration, an individual, say x_i is dominated by P_i individuals in the current population, then rank of x_i is given by $rank(x_i) = 1 + P_i$.

2. Fitness assignment

   (a) Sort the population in ascending order according to their ranks.

   (b) Assign fitness to individuals by interpolating the best (rank 1) to the worst (rank $\leq N$, N being population size)

   (c) Obtain a fitness score for each individual with a linearizing function.

   (d) Average the fitnesses of individuals with the same rank, so that all of them are sampled at the same rate.

   (e) Select the N individuals for reproduction.

Pareto-based approach: non-dominated sorting genetic algorithm (NSGA): This approach is based on several layers of classification of individuals.

1. The population is ranked on the basis of non-domination.

2. All non-dominated individuals are classified into one category (with a dummy fitness value, which is proportional to the population size to provide equal reproductive potential for the individual).

3. To maintain the diversity in the population, classified individuals are matched with their dummy fitness values.

4. Next another layer of non-dominated individual is processed.

5. The process continues until all individuals in the population are classified.

The flow diagram of non-dominated sorting genetic algorithm is shown in Fig. 2.20. Fitness sharing:

$$sh(d_{ij}) = \begin{cases} 1 - \left(\frac{d_{ij}}{T_{shared}}\right)^2 & \text{, if } d_{ij} < T_{shared} \\ 0 & \text{, otherwise} \end{cases}$$

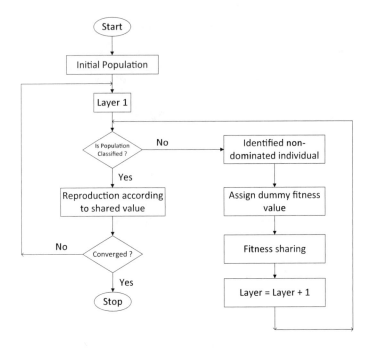

FIGURE 2.20: Non-dominated sorting genetic algorithm

Here d_{ij} is the distance (Euclidean distance) between the individuals i and j. T_{shared} is the minimum distance allowed between any two individuals to become members of a same niche.

$$d'_{f_i} = d_{f_i} \left[\sum_{j=1}^{N_{pop}} sh(d_{ij}) \right]$$

Here, d_{f_i} is the dummy fitness value assigned to individual i in the current layer and d'_{f_i} is the shared value. N_{pop} is the number of individuals in the population.

Pareto-based approach: niched Pareto genetic algorithm (NPGA)/Pareto dominant tournament selection

Input:

- N = Size of population
- $f = f(f_1, f_2, f_3, \cdots, f_k)$

Output:

    N' = number of individuals (dominants) to be selected.

1. $i = 1$ (first iteration).

2. Randomly select any two candidates C_1 and C_2.

3. Randomly select a comparison set (CS) of individuals from the current population. Let its size be N^* where $N^* = P\%N$ (P decided by programmer).

4. Check the dominance of C_1 and C_2 against each individual in CS.

5. If C_1 is dominated by CS, select C_2 as the winner. If C_2 is dominated by CS, select C_1 as the winner. If neither is dominated use do_sharing procedure to choose the winner.

6. If $i = N'$, exit because selection is done. If $i = i + 2$, return to step 2.

Procedure for do_sharing (C_1, C_2):

1. $j = 1$

2. Compute a normalized (Euclidean distance) measure with the individual x_j in the current population as follows.

$$d_{xj} = \sqrt{\sum_{i=1}^{k} \left(\frac{f_l^x - f_l^j}{f_l^U - f_l^L} \right)^2}$$

where f_l^i is the l-th objective function of the i-th individual and f_l^U and f_l^L denote the upper and lower values of the l-th objective function. Let σ_{share}= niched radius compute the following sharing value

$$sh(d_{xj}) = \begin{cases} 1 - \left(\frac{d_{xj}}{\sigma_{share}} \right)^2 & \text{, if } d_{xj} < \sigma_{share} \\ 0 & \text{, otherwise} \end{cases}$$

3. Set $j = j + 1$, if $j < N$, go to step 2 or else calculate niched count for the candidate as follows

$$n_1 = \sum_{j=1}^{N} sh(d_{xj})$$

4. Repeat steps 1 through 4 for C_2. Let the niched count for C_2 be n_2.

5. If $n_1 < n_2$; otherwise choose C_2 as the winner else C_1 as the winner.

2.4.2.1   GA-Based Multiobjective Optimization

Pareto-based approach

Multiobjective genetic algorithm (MOGA)
Non-dominated sorting genetic algorithm (NSGA)
Niched Pareto genetic algorithm (NPGA)

A decision vector u dominates v iff u is better than v with respect to (at least) one objective and not worse than v with respect to all other objectives. Here, $f_1(x^*) \leq f_2(x)$, $f_2(x^*) \leq f_1(x)$.

Evolutionary multiobjective optimization: Very often real world applications have several multiple conflicting objectives. To solve such a problem, evolutionary multiobjective optimization algorithms have been proposed.

What is multiobjective optimization? In a single objective optimization, the search space is well defined. On the other hand, when there is a contradicting objective to be optimized simultaneously, there is no longer a single optimal solution but rather a whole set of possible solutions of equivalent quality.

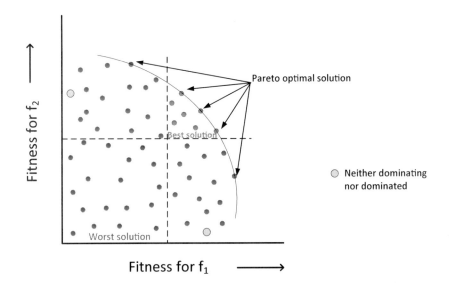

FIGURE 2.21: Pareto optimal

Pareto optimal: Let $\bar{F} = [f_1, f_2, \cdots, f_k]$ be a vector of objective function and $X = [x_1^i, x_2^i, \cdots, x_n^i]$ a design parameter. We say that at point $\bar{x}* \in \bar{X}$ is a Pareto optimal if for every

$$X_i \in F \; (i = 1, 2, \cdots, n)$$

either

$$\forall f_i \in \bar{F}[f_i(x_i) = f_i(x^*)], \; i = 1, 2, \cdots, k$$

or there is at least one $f_i \in \bar{F}$ such that

$$f_i(\bar{x}_i > f_i(x^*))$$

Figure 2.21 shows Pareto optima for different solutions.

Pareto front: A point x^* in $\bar{X}$ is a weakly non-dominated solution if there is no $\bar{X} \in X$ such that $f_i(\bar{X}) \leq f_i(X^*)$ for $i = 1, 2, \cdots, k$. A point x^* in $\bar{X}$ is a strongly non-dominated solution if there is no $\bar{X} \in X$ such that $f_i(\bar{X}) \leq f_i(X^*)$ for $i = 1, 2, \cdots, k$ and for at least one value of i, $f(\bar{x}) < f(x^*)$ (see Fig. 2.22).

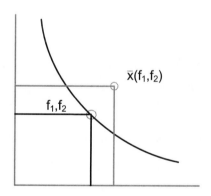

FIGURE 2.22: Pareto front

Pareto-based approaches:

1. Multiobjective genetic algorithm (MOGA)

2. Non-dominated sorting genetic algorithm (NSGA)

3. Niched-Pareto genetic algorithm (NPGA)

Pareto-based approaches: These approaches are also called evolutionary multiobjective optimization approaches. There are also a number of stochastic approaches such as simulated annealing, ant-colony optimization, swam particle optimization, and tabu search that could be used to generate the Pareto set. A good approach would be to generate the solution so that it yielded a good approximation with as little trade-off as possible.

Evolutionary multiobjective optimization: In general, evolutionary algorithms are characterized by a population of solution candidates and the reproduction process enables us to combine existing solutions to generate new solutions. Finally, the natural selection determines which individuals of the current population participate in the new population. A flowchart of the evolutionary multiobjective optimization is shown in Fig. 2.23. Two major prob-

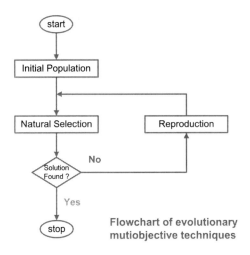

FIGURE 2.23: Flow chart of evolutionary multiobjective optimization

lems must be addressed when an evolutionary algorithm is applied to multiobjective optimization. How to accomplish fitness assignment and selection, respectively: guide the search toward the Pareto optimal set, and to maintain a diverse population in order to prevent premature convergence and achieve a well distributed tradeoff front.

Pareto-based approach: The basic idea is to use non-domination at ranking and selection to move a population toward the Pareto front. In other words, find the individuals that are Pareto non-dominated by the rest of the population. These individuals are then assigned the highest rank and eliminated from further contention. Next, another set of Pareto non-dominated individuals determined from the remaining population and are assigned the next highest rank. The process continues until the population is suitably ranked.

Issues: Devise an efficient algorithm to check for non-dominance (that is, Pareto set) in a set of feasible solutions. Traditional algorithms have serious performance degradation as the size of the population and the number of objectives increase. Next, we discuss different approaches addressing the above issues.

Multiobjective genetic algorithm (MOGA): There are two steps in MOGA.

1. Rank assignment

2. Fitness assignment

Rank assignment (MOGA): Fonseca and Fleming (1993) proposed this approach. They rank an individual corresponding to the number of chromosomes in the current population by which it is dominated. For example, an individual, say x_i of generation t, is dominated by a $P_i^{(t)}$ individual, in the current population. Then, x_i's current position in the population is given by

$$rank(x_i, t) = 1 + P_i^{(t)}$$

Note 1: All non-dominated individuals are assigned rank 1, while dominated individuals are penalized according to the population density of the corresponding region of the tradeoff surface.

Note 2: The rank of a certain individual corresponds to the number of chromosomes in the current population by which it is dominated. All non-dominated individuals are assigned the highest possible fitness value (all get the same fitness, such that they can be sampled at the same rate), while dominated ones are penalized according to the population density of the corresponding region to which they belong (i.e., fitness sharing is used to verify how crowded the region surrounding each individual is).

Note 3: This type of blocked fitness assignment is likely to produce a large selection procedure (i.e., degree to which the better individuals are selected) that might produce premature convergence. With this approach it is possible to evolve only a certain region of the tradeoff surface by combining Pareto dominance with partial reference in the form of a goal vector. If the basic ranking scheme remains unaltered as we perform a Pareto comparison of the individuals, the objective already achieved will not be selected.

Pareto-based approach: non-dominated sorting genetic algorithm (NSGA): This approach was proposed by Srinivas and Deb (1993) and is based on several layers of classification of individuals. Before selection, the population is ranked on the basis of non-domination.

1. All non-dominated individuals are classified into one category (with a dummy fitness value, which is proportional to the population size to provide equal reproductive potential for these individuals).

2. To maintain the diversity in the population, classified individuals are shared with their dummy fitness values.

3. Next, the group of classified individuals is ignored and another layer of non-dominated individual is processed.

The process continues until all individuals in the population are classified. A stochastic remainder proportionate selection is used in this approach.

Since individuals in the first iteration have maximum fitness values, they always get more copies than the rest of the population. This allows the search for non-dominated regions and results in quick convergence of the population toward such regions. NSGA efficiency lies in the way in which multiple objectives are reduced to a dummy fitness function using a non-dominated sorting procedure. With this approach, any number of objectives can be solved, and both maximization and minimization problems can be handled. A flowchart of the NSGA is shown in Fig. 2.24.

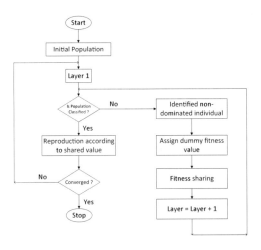

FIGURE 2.24: Non-dominated sorting genetic algorithm (NSGA)

The algorithm is similar to the simple GA except for the classification of non-dominated fronts and the sharing operation. The sharing in each front is achieved by calculating a sharing function value between two individuals in the same front as follows.

$$
sh(d_{ij}) = \begin{cases} 1 - \left(\frac{d_{ij}}{T_{shared}} \right)^2 & \text{, if } d_{ij} < T_{shared} \\ 0 & \text{, otherwise} \end{cases}
$$

Here d_{ij} is the distance (phenotype distance) between the individuals i and j, and T_{shared} is the maximum distance allowed between any two individuals to become a member of a same niche.

The parameter niche count $\sum_{j=1,i \neq j}^{N} sh(d_{ij})$ is calculated by adding the above sharing function values for all individuals in the current front. Finally, the shared fitness value of each individual is calculated by dividing the dummy fitness value by its niche count.

Use: Any number of objectives can be solved, Both maximization and minimization problems can be handled.

Fitness sharing: The fitness sharing techniques aim at encouraging the formation and maintenance of stable sub-populations (or niches). This is achieved by degrading the fitness value of points belonging to a same niche in solution space. Consequently, points that are very close will have their dummy fitness function values more degraded. The fitness value degradation of near individuals can be done using the following.

$$sh(d_{ij}) = \begin{cases} 1 - \left(\frac{d_{ij}}{T_{shared}}\right)^2 & \text{, if } d_{ij} < T_{shared} \\ 0 & \text{, otherwise} \end{cases}$$

Here parameter d_{ij} is the variable distance (Euclidean distance) between the individual i and j. T_{shared} is the maximum distance allowed between any two individuals to become members of a same niche.

$$d'_{f_i} = d_{f_i} \left[\sum_{j=1}^{N_{pop}} sh(d_{ij})\right]$$

Here, d_{f_i} is the dummy fitness value assigned to individual i in the current front and d'_{f_i} is the shared value. N_{pop} is the number of individual in the population.

Integration in soft computing: We have the characteristics of individual soft computing tools. While their different applications have been demonstrated on various real-life problems, there has been a great deal of research addressing synergistic integration of these individual tools since the late 1980s. The objective is to obtain application-specific benefits superior to those of the existing ones. Such integrations include neuro-fuzzy computing [2], fuzzy-genetic computing [4], and rough-fuzzy computing [5]. Rough-fuzzy computing provides a stronger paradigm for handling uncertainties arising from overlapping concepts and granularity in the domain of discourse. Therefore, soft computing approaches have also been widely used in solving different problems in sensor networks.

Bibliography

[1] S. K. Pal and D. Dutta Majumder. *Fuzzy Mathematical Approach to Pattern Recognition*. John Wiley & Sons, New York, 1986.

[2] S. K. Pal and S. Mitra. *Neuro-Fuzzy Pattern Recognition: Methods in Soft Computing.* John Wiley & Sons, New York, 1999.

[3] S. Pal, S. Oechsner, B. Bellalta, and M. Oliver. Performance optimization of multiple interconnected heterogeneous sensor networks via collaborative information sharing. *Journal of Ambient Intelligence and Smart Environments*, 5(4):403–413, 2013.

[4] S. Bandyopadhyay and S. K. Pal. *Classification and Learning Using Genetic Algorithms.* Applications in Bioinformatics and Web Intelligence Series. Springer, Heidelberg, 2007.

[5] P. Maji and S. K. Pal. *Rough-Fuzzy Pattern Recognition: Application in Bioinformatics and Medical Imaging.* Bioinformatics, Computational Intelligence and Engineering Series. Wiley–IEEE Computer Society Press, 2012.

[6] G. Rozenberg, T. Back, and J. N. Kok, Editors. *Handbook of Natural Computing.* Springer, Heidelberg, 2012.

[7] A. E. Eiben and J. E. Smith. *Introduction to Evolutionary Computing.* Springer, Heidelberg, 2003.

Part II

Fundamental Topics

Chapter 3

Evolution of Soft Computing Applications in Sensor Networks

Priyanka Sharma

International Institute of Information Technology

Garimella Rammurthy

International Institute of Information Technology

3.1   Introduction

In the past decade, an evolution has been witnessed in some of the major information processing techniques which are not all necessarily analytical types. Specifically fuzzy logic, artificial neural networks, and evolutionary algorithms have witnessed major transformations. The key ideas of these algorithms have their roots in the 1960s. As years passed, these algorithms developed based on the needs to solve the problems. In the 1990s the need of integrating different algorithms came into existence. Most importantly, the need was to develop a more powerful tool which would combine strengths of all the existing algorithms to solve high complexity problems. At this point of time, soft computing came into existence and was proposed as a synergically fused algorithms tool which promised to yield innovative promising solutions in the future.

Prior to 1994, soft computing was considered an isolated concept. However, over time soft computing can be seen from the perspective of a series of techniques which can help in dealing with real-time problems based on intelligence, common sense, and common analogies. From this perspective, soft computing can be understood as a method of problem resolution which makes use of approximate reasoning and optimization of problems. When it comes to intelligent systems, soft computing would form the theoretical basis.

This also helps in understanding the difference between artificial intelligence and intelligent systems. Artificial intelligence is based on hard computing. On the other hand, intelligent systems are based on soft computing. Initially, when soft computing came into existence it was mostly related to fuzzy logic and fuzzy systems. However, over time the methodologies evolved and other algorithms have also become a part of soft computing. Initially it was about approximate reasoning and its related methods. It has been mainly due to the constant attention paid to the development of these techniques that allowed soft computing approaches to be applied to different fields and industries such as telecommunications, computer science, sensor networks, automation, and many others.

3.2 Overview of Soft Computing and its Applications

3.2.1 Definition of Soft Computing

"What is soft computing" is one of the important questions to be asked and answered before starting to work on soft computing and its techniques. The field of soft computing is relatively new and has been evolving rapidly and this contributes toward the difficulty of answering this question in a precise manner. The definition of soft computing has been made in different ways: listing its technologies, defining properties of these technologies, and comparing it with different concepts and defining its use. This implies there are multiple definitions from different perspectives.

In this section, we provide an analysis of soft computing definitions from varied dimensions such as unification of different techniques, properties and others. Before we begin, to discuss definitions, we will first present the idea of soft computing as it was in the beginning. Lotfi Zadeh is referred to as the father of soft computing. When defining the concept initially in 1990, the hammer principle came across Zadeh's mind. It states that when what you have is a hammer, everything seems to be a nail. Similarly in science, when we are committed to a particular type of methodology as the best answer which exists. However, soft computing is different and it rejects the hammer principle in science. When elaborating on soft computing definitions this key aspect needs to be kept in mind (Magdalena, 2010).

Soft computing is an evolving concept and is not homogeneous. Distinct methods confirming its guiding principle come together to form the concept. There have been changes in the key elements of soft computing applications which relate to the evolving nature of soft computing. The evolving nature of soft computing is defined as amalgamation of different techniques which do not have fixed and clear borders. The evolving characteristic of soft computing leads to a number of criticisms regarding a certain technique in soft computing. At this point, it is important to understand that soft computing is not just a mixture of different techniques. It is a technique which incorporates a family of closely interacting techniques. The term hybridization can then be considered important when it comes to soft computing.

3.2.2 Soft Computing and Hybridization

The main contribution of soft computing is the hybrid approach it has given to the community since its development. Consider the example of neuro fuzzy systems, which is an important hybrid approach provided by soft computing. Neuro fuzzy systems combine fuzzy logic and neural network elements. Hence, the element of hybridization is present. Figure 3.1 presents the elements of hybridization in soft computing. Another way of defining soft computing is

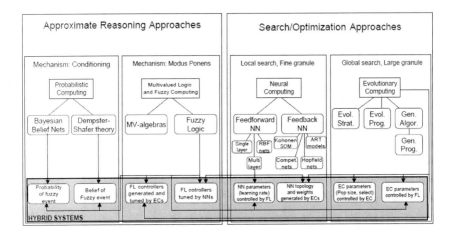

FIGURE 3.1: Elements of hybridization in soft computing

through its essential properties. It can be defined as a computing process which involves precision and allows the precision to change the result (either increase or decrease it). In such a scenario, the computing process is said to be included in the field of soft computing approaches. Hard computing approaches are computing processes which are not flexible and are demanding computationally. The demand of computationally approaching the problem in hard computing problems made them difficult to implement in different problem scenarios. Hence, soft computing approaches were developed and soft computing is also defined as the complement of hard computing problems.

Hard computing approaches make use of a definite model to solve problems. However, soft computing approaches derive the model from the data set which is available through the problem. Another definition of soft computing calls it an approach which helps in handling uncertainty and imprecision.

When working on a problem, there are times when the result may be noisy or the data set is imprecise or the information available. In such scenarios, it would be difficult to get absolute results. However, soft computing allows us to incorporate the uncomfortable characteristics, makes use of approximate reasoning and does computing using perceptions. An important point is to note that soft computing approaches do not target the problems with uncertainty. Instead, they solve with equal efficiency. The inclusion of uncertainty and noise, makes soft computing applicable to a large number of real-world problems.

3.2.3　Key Elements of Soft Computing

The main components of soft computing are neural computing, genetic algorithms and fuzzy logic. These components function as complementary techniques under the soft computing group. Different components of soft computing are derived from natural phenomena. For instance, the idea of neural computing is based on how the brain works, the genetic algorithm is derived from Darwinian evolution and fuzzy logic comes from uncertain nature of human speech. Now we will discuss the different components in detail.

1. Fuzzy Logic – When communicating with people in general, humans use a number of qualitative words. There is a lot of imprecision and uncertainty in human speech. For instance, the phrases "very young" and "a little far" provide a hint of what the person is saying. Fuzzy systems comes from stiff Boolean logic. Rather, they are a generalization of the same. They make use of fuzzy sets which come from crisp sets. Crisp sets come from classical set theory, as per which an element can either be a member of the set or not. However, in fuzzy systems the membership of an element is defined as degree of membership. The value of membership in case of crisp sets is either 1 or 0. On the other hand, in case of fuzzy systems, the membership is in the range of 0 to 1. In general, a fuzzy system is made up of fuzzy rule base, fuzzification module, an inference engine, and defuzzification module. The roles of different modules are as follows:

   - Fuzzification module – It pre-processes the input values which have been provided to the system.

   - Inference engine – The results provided by the fuzzification module along with the rules defined in the fuzzy rule base are used and results are inferred by the inference engine.

   - Defuzzification module – This module provides the final output of the complete fuzzy expert system.

   The motivation of development of fuzzy logic was to fulfill the need for a framework which dealt with concepts while handling uncertainty. The perspective of exact reasoning is as a limiting case of approximate reasoning. All the measurements done in fuzzy logic are in the form of degrees. Knowledge is a combination of flexible, fuzzy constraints on a group of variables. The flexible constraints are propagated to produce inference in a fuzzy system. A system which involves logic at some level can be fuzzified. The main features of fuzzy systems (which contribute toward its efficient performance) are its ability to deal with approximate reasoning specifically where a mathematical model is absent or cannot be derived easily. The estimated values are taken into consideration using which decisions are made in a fuzzy system (Chaturvedi, 2008).

2. Evolutionary Algorithms – Human genetic concepts are simple, unrestricted by domain, make use of probabilistic approach and are powerful. These characteristics of human nature are useful in evolutionary

algorithms. The models of evolutionary algorithms, make use of natural selection which involves recombination and mutation of different characteristics.

3. Genetic Algorithms – Genetic algorithms were developed on the principle of genetics and make use of a population of individuals referred to as the potential solution in genetics. The desirable optimal solution is obtained when the algorithm converges. Genetic algorithms can be combined with other sets of algorithms to solve a domain-specific problem. They can be implemented at the machine level to obtain intrusions at the real time mode. The most important dimensions of genetic algorithms are defining objective function, defining and implementing the genetic representation and defining and implementing genetic operators. These dimensions form the basic components, and variations can be designed to improve performance of genetic algorithms.

Suppose we are given a function; using genetic algorithms we can optimize this function by making use of random search. Problems involving optimization along with absence of a priori information can be best approached using genetic algorithms. If we look from the perspective of an optimization method, the efficiency of genetic algorithms increases. From the parent generation, genetic algorithms create a child generation based on a defined set of rules which mimic the genetic reproduction. In case of these algorithms, randomness is an important factor. This is mainly because the selection of parents is done randomly. However, the parents who are best, have an increased probability of getting selected in comparison to other parent populations. Similarly, the number of genes is a random selection. The components discussed above when combined and used in a hybrid manner contribute toward the efficiency of soft computing algorithms.

3.2.4 Significance of Soft Computing

Since the development of fuzzy logic in the 1990s, different methods based on fuzzy sets have become an important part of all areas of research. There has been a lot of innovation as a result of the implementation of fuzzy logic. Besides research, the idea has been extended to implementation in daily life scenarios of health, home, banking and education. Also, soft computing applications have helped us over time to understand and solve problems of high complexity. They help in solving both comprehension and experimental aspects of different problems. The significance of soft computing is not limited to fuzzy logic and its applications. Evolutionary algorithms have proved to be of high significance in finding solutions to multiple problems. The important aspect of evolutionary algorithms is that they can be applied to different applications and can be expected to provide efficient solutions to the intelligent systems. Over the years, soft computing has grown and added different algorithms which have contributed toward its significance.

The different components of soft computing algorithms can be used with each other to solve problems in an effective manner. In some problems their combination proves to be more effective than their exclusive functionality. Consider the example of neurofuzzy systems. The usability of these systems increased with their implementation in different products (such as air conditioners, washing machines, photocopiers, camcorders). The implementation of soft computing approaches in industrial products and consumer products eventually resulted in a high machine intelligence quotient. The high machine intelligence quotient contributes toward the increase in applications of soft computing both in terms of number and variety. The logical framework of soft computing means that the university students in future should be exposed to fuzzy logic, the basic components of soft computing and also to the related methodologies of soft computing. For instance, Berkeley Initiative on soft computing delivers courses to students with interest in soft computing. Such initiatives contribute indirectly toward the development of machines with high intelligence and critical decision making power.

3.2.5   Goals of Soft Computing

The aim of soft computing is to explore the imprecision which can be accommodated in a particular problem. It also exploits uncertainty and approximate reasoning. The main reason behind exploring different aspects of a problem is to understand its resemblance to human brain decision making. The primary goal is to develop machines which are intelligent and to solve problems which are nonlinear and not mathematically modeled. Over time the applications of soft computing have shown two advantages. First, the nonlinear problems in which mathematical models are not available can be solved now with the help of soft computing algorithms. Further, the knowledge imparted by the human brain is now understood as different forms such as recognition, cognition, learning, understanding and others. These functions are now related and understood from the perspective of computing. Intelligent systems can be built to be autonomous and tune themselves.

3.2.6   Applications of Soft Computing

Soft computing has found applications in many diverse fields. It is applied to solve business, industrial application, engineering problems, and others. For instance, in engineering problems soft computing finds applications in restoration of damaged road networks and maintenance planning of bridge structures. In other fields as well, soft computing is applied when the need is for an intelligent system that can learn or predict.

3.3   Sensor Networks and Soft Computing Applications

3.3.1   Intrusion Detection

Sensor networks have been widely applied to different systems. Their wide deployment always raises the question of their safety and security. Cryptography has been generally used to make them secure from any type of unauthorized access. Cryptography, intrusion detection mechanisms have also been implemented for making WSNs secure. However, we still need new techniques which would help in making WSNs systems more secure. This is where soft computing is needed. The complementary nature of techniques of soft computing could contribute toward development of intrusion detection mechanisms which would protect them from attacks or discrepancies. For instance, classifiers based on fuzzy rule, techniques which come under genetic programming, support vector machines and ant colony systems can be used for developing intrusion detection for WSNs. A general cryptography technique or an intrusion detection system would provide a defined method of protecting WSNs. The intelligence quotient provided by soft computing approaches would help in development of a fine protection system which would be difficult to breach as well. An important aspect of soft computing approaches which can be useful in WSNs, is the ability to identify or at least track behavioral patterns performed in a deliberate manner and those which have been performed involuntarily. The self-organizational principle and probability values help in the development of these safe and secure systems for WSNs. Hence, the deployment of soft computing approaches is needed in WSNs for intrusion detection.

3.3.2   Localization

In case of WSNs, it is important that the sensor nodes are localized. This is mainly because the general pattern of placement of sensor nodes is random scattering. A region of interest is selected and sensor nodes are placed. The connection made between sensor nodes and the network is independent. The need for localization is in data routing, reduction in energy consumption and handling data which is dependent on location. A large number of WSN applications demand localization from the sensor nodes. In such cases, where localization is important, WSNs can be accompanied by soft computing approaches such as a neural network. The network can be trained using distant noises. The neural network can further compensate for the noise and help in finding the exact location of the sensor node. Accuracy can be achieved in this scenario. Further, other components of soft computing can also be applied with the same idea. Another important aspect to consider when implementing such systems is to minimize utilization of resources. The overhead should remain as low as possible. The need is to accurately localize different sensor nodes in a wireless sensor network.

3.3.3   Optimization of WSNs

As a result of development of micro electro mechanical systems, WSNs have also witnessed tremendous development and popularity. WSNs connect to the virtual world through sensor nodes. Energy remains a constraint for sensor nodes and their deployment in real-time scenarios. Since they are battery operated, it becomes difficult to provide them with energy or power at all times. Therefore, designing WSNs is a difficult task concerning this particular issue. Cross layer optimization is one of the approaches which can be used for increasing the lifetime of a network and improve its functioning. All the layers can be allowed to interact with each other. In WSN, the pattern of deployment is random which brings into picture the important aspect of coverage area. For all the nodes it is important that the effective distance is adequate. This would also have a potential influence on the throughput and the functioning of the WSNs. To design WSNs in an optimal manner, it is important to take into consideration energy issues of the sensor nodes and the coverage area of the same. The idea is to use soft computing approaches to provide the WSNs with optimal design parameters. Some of the very specific techniques of soft computing such as bacteria foraging algorithm, swarm intelligence, and others can be put to use in order to design a WSN which provides high performance and throughput, keeping in mind energy constraints and coverage area issues. By making use of soft computing approaches, maximum coverage can be obtained by deploying minimum sensor nodes in the form of a cluster by installing a transceiver between them to help in communicating with all the sensors. For instance, consider a bacteria foraging algorithm, according to which the animals whose strategies are weak are eliminated. To be precise, bacteria foraging natural selection eliminates the animals whose strategies of foraging are weak. It favors the propagation of animals with strong foraging strategies. The same idea can be put to use in sensor networks as it would help in designing a strong and optimum WSN (Chong and Kumar, 2003).

3.3.4   Applied Systems

Machine Health Monitoring and Structure Monitoring: WSNs are deployed in different fields among which is health. Wireless sensor networks are applied to a machine health monitor consisting of both hardware and software and also possess a control component for communication.

A health monitor can be implemented in wired or wireless form. The earliest forms were wired. The absence of wires in later wireless systems reduced both machine cost and operational overhead. Energy is provided by batteries and power is consistent. A sensor in wireless system monitors the operating condition of the machine. The recorded maintenance information is sent to LabVIEW[1].

The graphical interface of LabVIEW accepts results and notifies maintenance staff of disturbances. LabVIEW employs an enery efficient protocol in transmission of information to prevent data collision. The question is how soft computing enters the picture.

Notification via graphical interface takes time. Approaches such as neural networks can be effective for notifying maintenance staff quickly. Use of soft computing makes a system operate faster. It may also allow it to predict future disturbances to maintenance people and enable them to be proactive in solving problems promptly.

WSNs are also deployed for structure monitoring. They can inspect damage to structures such as bridges, buildings, and ships using an excitation-and-response scenario. The structure's response to excitation helps determine the location and extent of damage. An example of such a system is Wisden. Soft computing approaches are needed to increase the machine intelligent quotients of such systems. Soft computing may be utilized to measure other parameters after a period of learning and will ultimately reduce damages to structures through effective monitoring.

Environmental Monitoring: Environment has become an important concern recently. The sensing ability of WSNs have found application in environmental monitoring and may be extended to other areas of research as well. For instance, WSNs have been deployed to study the behavior of birds in the United States. The cluster formation of burrows was discovered and sensors are deployed for studying these clusters. The aim was to minimize the disturbances from humans. To transmit information, the communication link uses multiple hops and sends and receives information through different contact points. The behavior of animals and other aspects of environment can also be measured using WSNs, for example, humidity, temperature, pollution, and others. The same idea of monitoring can also be extended to the agricultural field.

The approaches incorporated in soft computing are derived from patterns of human behavior. It considers the human patterns of functioning and then predicts the results. The inclusion of soft computing in understanding the environment would help from a very important perspective. The tendency of extracting from natural behavior can be correlated and then deployed to explain animal behavior, agricultural, or the environment. As we discussed above, multiple hop transmission link is needed to get results. The use of soft computing approaches can reduce overhead and produce simpler and cost effective results at the same time.

---

[1]LabVIEW is a system design platform and development environment from National Instruments for data acquisition, instrument control, and industrial automation.

Health Care: WSNs have found application in health care. WSNs are used to measure and monitor patient statistics and notify staff in times of emergency. They also help in detecting certain symptoms which may be hard to detect otherwise. For instance, brain tumours can be identified by monitoring brain activity through sensors. Since the application of wireless sensor networks is common in health care, it is important that the measurements are fast and highly accurate. Soft computing algorithms are needed in such systems for several reasons. First is to support the wireless sensor network. There may occur a time when a sensor node is falling short of power or energy or has completely shut down. In such scenarios, soft computing approaches can support the network and help predict or identify when a failure has occurred. It can also provide intelligence to the sensor and further enhance the accuracy of these systems in health care.

Another benefit from soft computing is monitoring and prediction. Patient statistics are monitored at present, but the sensor networks of the future will allow the predictions to be made based on statistics. Of course, accuracy and reliability are issues that must be resolved along with appropriate learning protocols. Research is ongoing to find uses for sensor networks in other areas of health care. The intelligence and sensitivity quotient of soft computing when applied to sensor networks would only enhance the system. Therefore, the need is to develop soft computing along with sensor networks for the health sector.

Fire Rescue, Humanitarian Search and Rescue Systems: Another application of wireless sensor networks which demands accuracy and high speed response is fire rescue. The system to be designed for fire rescue demands real-time monitoring, allocation of resources, and scheduling done in a smart and intelligent manner. These systems require soft computing approaches to be included in their deployment. A similar pattern applies to humanitarian search and rescue systems.

We will consider certain scenarios to validate the need of soft computing approaches for the wireless sensor networks. First consider fire rescue. In large scale operations, for example in cities, WSNs must be designed to handle certain functions such as detecting fires and notifying the rescue crews of the extent of the fire and people who need rescue. A neural network or machine learning method would provide help in those areas and also furnish helpful information to rescuers, for example, increase in temperature. Reinforcement learning or iterative learning could enable a system to identify where fire victims could be found.

After an earthquake, sensor networks could help find missing victims after they learn to follow certain parameters. A system could use past searches to predict where victims could be found. The efficiency of such systems could be increased by using a set of fuzzy rules to help them deal with uncertainties. Fuzzy logic would be effective for dealing with random scenarios because no particular model is associated with it.

Traffic and Transportation Management: Consider a WSN for avoiding collisions at road intersections. Traffic management is an important application area of WSN. For avoiding collisions every car is provided with a radio module to show where other cars are in the surrounding area. The need here is to improve these systems. Since sensors are short ranged and limited in functionality, the car collisions can not be stopped entirely. Therefore, soft computing approaches can be implemented in traffic and transportation management WSNs. A central system can be developed using WSNs and soft computing approaches to warn of possible collisions in the time frame of a human reflex. There are other possibilities that need to be explored. Information would be collected to predict the possible trajectories of the cars in the surrounding areas. This would help the drivers be aware of the path on which other cars are already driving. Vehicles carrying goods (such as medicines, automobiles, explosives, or other products) can be assisted by WSNs and soft computing approaches before any loss happens or any damage to the vehicle or passengers takes place. A high machine intelligent quotient in a system with soft computing and WSNs would help resolve different types of situations on the road.

Now, we understand the need of implementing soft computing approaches to varied applications of WSNs. A lot of research still needs to be done to get effective results to further improve these systems and make their functioning even better. The ability of a system to judge, predict, make decisions, and control scenarios would greatly develop existing technologies.

3.4   Case Studies: Varied Implementation of Soft Computing Applications in Sensor Networks

3.4.1   Case Study 1: Smart Node Model for Wireless Sensor Network

The main component of a smart node is a fuzzy engine composed of three sub-components: knowledge base (a set of fuzzy rules), fuzzification, and defuzzification, which changes measurements in numerical form to linguistic form and vice versa. A smart node can perform approximation on any function of input variables, e.g., when it is difficult to quantify certain variables, they can be calculated by smart nodes with the help of a set of special rules in the knowledge base. Similarly these rules can be developed for other functionalities such as data fusion and aggregation, wireless routing, and others. The knowledge base of a smart node can be developed with the help of knowledge gained through supervised neural network learning. The knowledge obtained would help in improving resource utilization and efficiency of energy utilization, for instance, data caching (done for a particular application) happens in intermediate node (smart node) through the MeshNetics TM platform.

MeshNetics was developed for enhancement of M2M applications. It comprises software components, algorithms, hardware designs, and solutions. Through addition of expert systems, the needs and requirements of end users are met as the expert systems help in controlling, monitoring, and getting results from different applications.

An entirely new era of wireless sensor networks has come into existence as a result of the embedded expert system with complex algorithms enabling advanced tasks to happen quickly and efficiently using neural networks and fuzzy logic algorithms. These expert systems have also helped business to get rid of costly and static wired systems. The embedded expert systems are put on a base which permits a hybrid distributive system in the form of smart nodes in wireless sensor networks. This helps in data collection, communication and also in fusion of data along with data aggregation. Meshnetics TM is a smart node platform, consisting of a smart engine driven by a number of variables, such as pattern of rules, parameters, member functions, and others. These variables can be sent from one wireless sensor network to another. The user is provided with the ability to define these transmissions. A smart node can also define a transmission on its own. For instance, in the process of goal tracking, the transmissions can be achieved with the help of smart nodes working in dependence with other smart nodes. For making intelligent decisions in WSN, the software environment of smart nodes is put to use. This decision making ability is associated with different research areas in embedded computer systems (with the property of being power aware) network architectures based on applications, operating systems which are restricted in resources, distributed algorithms, aggregation, and processing.

3.4.1.1   Smart Node in WSN Control

When making use of knowledge for the purpose of controlling wireless sensor networks the paradigm of the active network is taken into consideration. A number of aspects can be realized with the help of smart nodes:

- Routing algorithms
- Optimal control of power consumption
- Data traffic control
- QoS control

3.4.2   Using SN for Fuzzy Data Base Designing

There exists a high level of similarity between sensor networks and distributed database systems. Distributed database systems store environmental data and provide aperiodic responses. In sensor networks, data requests can be registered a priori, for the transmission of data to take place only when it is needed.

This would prevent any unnecessary transmissions. This feature is similar to the traditional database in which data is searched through queries which filter data. The filtering handles large volumes of data. Similarly, fuzzy queries also contribute toward reducing the volume of raw data.

3.4.2.1 Using SN for Data Aggregation and Fusion

There might be significant power loss and possible collisions if all the raw data is forwarded to the base stations. To minimize unnecessary transmission of data, intermediate nodes or neighboring nodes can work together for filtering and aggregation of data arriving at the destination. Data association, effect estimation, and identity fusion are the goal-oriented methods [1].

3.4.3 Case Study 2: Soft Computing Protocols for WSN Routing

Energy is an important aspect of WSNs. Both performance and lifetime of a network are influenced by utilization of energy. Due to its importance, the attention paid to energy has increased in recent times. As a result, power management approaches are being sought. Keeping in mind the constraints of WSNs and the scarcity of energy in the sensors, it is important that routing is done in the most effective manner possible so that the energy use is balanced among all the nodes. This would contribute toward increasing the lifetime of the network and also ensure the coverage of the network. To increase the effectiveness of sensor networks, the need is for smart techniques which intelligently handle different operations of the network. The paradigms defined in soft computing contribute to studying and examining optimization of routing in WSNs along with ensuring energy efficiency. These applications also take into consideration the design and deployment issues of WSNs [5]. The flexibility of soft computing approaches combined with the usability of WSNs have increased the usability WSNs.

3.4.3.1 Reinforcement Learning

In wireless routing, reinforcement learning proved to be greatly effective. The placement of sensor nodes in the network along with the distances at which they are placed from the sink nodes has a great influence on the amount of energy which is consumed by each node. For each node, it is crucial to ensure reasonable energy consumption along with algorithms for optimization. This helps in improvement of performance and preservation of energy within network. The idea behind reinforcement learning is to provide solutions for problems related to routing approaches in sensor networks along with minimum computational requirements. The adaptive nature of sensor networks helps them save energy in unnecessary communications and also help in balancing the node in a dynamic manner. Research in this field has suggested an important innovation for this purpose which is the Q-learning algorithm.

Based on the fact that the nodes which are directly connected have a lot of information related to energy of the network, to incorporate reinforcement learning in wireless sensor networks. Self-learning by each node further contributes to improvement of efficiency and overall performance.

3.4.3.2   Swarm Intelligence

Swarm intelligence has shown great compatibility with routing in wireless sensor networks. Due to the match of swarm intelligence and WSNs, a number of efficient techniques of routing can be derived. Swarm intelligence takes into account the group behavior of insects which shows dynamic nature and is distributed in a decentralized system. The basic idea of swarm intelligence is to optimize distribution of uncontrolled systems, based on which routing techniques could be derived. For instance, the shortest path in an ant colony could serve as a derivation of routing protocol for WSNs.

The swarming behaviour of ants led to development of a number of innovations, for example, the basic ant-based routing algorithm, the sensor driven and cost aware routing algorithm, flooded piggybacked ant routing, and others. The goal of all these techniques is to find the optimal shortest path between source and destination nodes while considering energy conservation to improve network lifetimes.

Energy is an issue when propagation from one node to another occurs along a defined transmission path. After data is delivered, the amount of energy consumed must be updated during the return journey along the same path. Swarm intelligence exhibits great flexibility in WSN routing but it faces the limitation of adding to existing traffic. Furthermore, the forward and backward routing are not flexible for WSNs although they keep overhead low.

3.4.3.3   Evolutionary Algorithms

Evolutionary algorithms have not proven optimum for WSN routing. This is mainly because the lifetime of a WSN is dependent on energy consumption of the nodes, along with selection of an optimized path between source and the sink which is the main challenge. For finding the shortest path, we need to calculate the cost function of the path between source and sink. The best fit of evolutionary algorithms is when the problems are unconstrained. Research in evolutionary algorithms found an optimal path on the basis of genetic algorithms which also optimizes the cost function.

A very basic genetic algorithm has been put to use. An initial energy level is set equal for all the nodes. Using an improvised version of the elitism concept, the shortest route (which also possess energy efficient ability) is nominated. The use of this model would provide survival of the best individuals in the next generation along with important improvements of convergence. For improving network lifetime, a defined threshold is applied for power on each node. The best energy nodes are then replaced by weak energy nodes.

On the basis of probability calculation, if the energy of the node is less than threshold, the node will be eliminated from the network. There would be insertion of new nodes for maintenance of diversity. Because genetic algorithms have a strong base for computational algorithms, it is difficult to consider them efficient for WSN routing. The learning phase is offline which also required high number of calculations for a routing tree. Fitness tests are performed to select the best population. A problem function needs to be defined in WSNs for calculation of total cost function. The successive implementation of selection, crossover, and mutation helps in optimal selection. The total power and network lifetime depend on the collected energy among source and sink. The relationship is inversely proportional which means, that power and lifetime will increase if there is a decrease in the energy collected between source and sink.

3.4.3.4   Fuzzy Logic

Since the environment of WSNs involves uncertainty, decision making needs to be done appropriately. Procedures contributing to deployment of routing and enhancement of total lifetime of network are needed. These procedures also need to be flexible for accommodating variations. The implementation of fuzzy logic would help in providing flexibility when dealing with imprecision in the network and it will also help in avoiding complex mathematical modeling. A number of algorithms have been proposed in fuzzy logic with the same idea: fuzzy dynamic power control algorithm, improvement of LEACH protocol by fuzzy logic, and others.

3.4.3.5   Neural Networks

Artificial neural networks are algorithms which can perform learning and mapping of complex relations. The basis of mapping is supervised learning that functions in multiple environments. The application of neural network algorithms from the perspective of wireless sensor networks provides in depth understanding of many issues of this particular environment.

The architectures of neural networks and WSNs are analogous. The connections of the nodes in both type of networks are the same. The traditional signal processing algorithms of WSNs can be replaced by the computational algorithms of artificial neural networks (ANNs). The similarities help reduce resource use and promote efficient implementation. The future of ANNs in relation to WSNs is secure because of their compatibility.

The applications of ANNs to WSNs would predict and reduce the amounts of energy required for routing. The need is to increase the scalability of sensor networks along with decreasing the energy consumed. ANNs could provide nodes with sensing abilities. This would help reduce energy needs through measurement and prediction.

However, neural networks cannot accommodate certain challenges. For example, the frequently changing behaviors of WSNs are difficult for neural networks to understand. The overhead is in offline learning and calculations that cannot be adapted easily to WSNs.

3.4.3.6 Artificial Immune Systems

Artificial immune systems are not widely applied to wireless sensor networks. Artificial immune systems demonstrated the ability of applying clustering and selection of cluster heads in a network. This provides a new research potential for artificial immune systems in wireless sensor networks. The algorithms and paradigms defined in artificial immune systems have been implemented in different ways in research. However, a lot of work still needs to be done to explain the operation in a better way. The implementation of artificial immune systems can be broadened to different applications of wireless sensor networks.

3.5 Future Scope of Soft Computing Applications in Sensor Networks

The future of sensor networks appears bright for implementation of soft computing applications. The components of soft computing exclusively or together can be used to solve or further simplify the problems of sensor networks such as data aggregation, network lifetime, energy consumption and energy utilization, information processing, and others. A lot of research is in progress but the implementation of both soft computing and sensor networks is limited at present. It is important to take the research and its results further and implement sensor networks while making use of soft computing applications.

3.6 Summary

The definition of soft computing is evolving and the approach can be defined from different perspectives. In general, the approach utilizes basic components which can be applied to many fields for developing intelligent systems. The main aim of these approaches is to increase the machine intelligence quotient of different systems. Witnessing the developments taking place in wireless technology, wireless sensor networks have found application in large real-time scenarios. However, these wireless sensor networks still face limitations due to energy constraints and other factors. The implementation of soft computing applications in wireless sensor networks adds to network lifetime and improves processing abilities in different scenarios.

Bibliography

[1] A. N. Averkin, A. G. Belenki, and G. Zubkov, Soft Computing in Wireless Sensors Networks, in *Proceedings of Conference of European Society for Fuzzy Logic and Technology*, pp. 387–390, 2007.

[2] D. K. Chaturvedi, *Soft Computing: Techniques and Its Applications in Electrical Engineering*, Springer, 2008.

[3] C. Y. Chong and S. P. Kumar, Sensor Networks: Evolution, Opportunities, and Challenges, in *Proceedings of the IEEE*, Vol. 91, 2003.

[4] L. Magdalena, What is Soft Computing? Revisiting Possible Answers, *International Journal of Computational Intelligence Systems*, Vol. 3, pp. 148–159, 2010.

[5] H. N. Elmahdy, M. Sharawi, I. Soriet, and E. Emary, Routing Wireless Sensor Networks Based on Soft Computing Paradigms: Survey, *International Journal on Soft Computing, Artificial Intelligence and Applications*, vol. 2, pp. 1–15, 2013.

[6] J. Brignell, The future of intelligent sensors: a problem of technology or ethics?, *Sensors and Actuators A*, Vol. 56, pp. 11-15, 1996.

[7] X. Cui, L. U. Hardin, T. Ragade, and A. Elmaghraby, A swarm-based fuzzy logic control mobile sensor network for hazardous contaminants localization, in *Proc. of the IEEE International Conference on Mobile Ad-hoc and Sensor Systems*, 2004.

[8] A. Kulakov and D. Davcev, Tracking of unusual events in wireless sensor networks based on artificial neural-networks algorithms, in *Proc. of International Conference on Information Technology: Coding and Computing*, 2005.

[9] G. Molina and E. Alba, Location discovery in Wireless Sensor Networks using metaheuristics, *Applied Soft Computing*, Vol. 11, pp. 1223–1240, 2011.

[10] W. V. Norden, Intelligent task scheduling in sensor networks, in *Proceedings of International Conference on Information Fusion*, 2005.

Chapter 4

Soft Computing Approaches in WSN Routing Protocols

Konjeti Viswanadh

International Institute of Information Technology

Munjuluri Srikanth

International Institute of Information Technology

Garimella Rammurthy

International Institute of Information Technology

4.1 Introduction

Standard telecommunication network texts define "network planning and design" as an iterative process, encompassing topological design, network-synthesis, and network realization, and is aimed at ensuring that a new telecommunications network or service meets the needs of the subscriber and operator. It involves a great amount of traffic engineering in it. Very soon, the design of a WSN is going to be as complicated as designing a telecommunication network. The huge increase in mobile phone use, the concept of the "Internet of Things," and advancements in VLSI technologies that led to development of miniature devices will create a need for innovative designs of WSNs.

Their designs will become more complex because of (i) their inherent challenges such as energy constraints and (ii) the abilities of networks to scale up to demands. The first challenge can be dealt with; many textbooks have covered WSN designs. Scalability is a challenge because installation of dedicated WSNs is not possible. Every node must handle multiple networks simultaneously, for example, a mobile phone may be part of a telecommunications network and part of a WSN.

In order to make the WSN design simpler, soft computing paradigms can be brought into play. Soft computing techniques were developed for fields, but surprisingly find many applications in WSNs. Almost every WSN design decision, especially routing decisions, can be assisted by the soft computing approach, discussed in the later sections in this chapter.

This chapter is organized into six sections. Section 4.2 presents an overview of routing protocols. Section 4.3 discusses challenges. Section 4.4 provides a comprehensive overview of all soft computing paradigms. Section 4.5 describes future research directions and Section 4.6 covers conclusions.

4.2   Routing Protocols: Classification and Overview

Due to the limitations on energy resources in a WSN, prolonging the network lifetime is considered a challenge. Although network lifetime is affected by the limitations of the battery powered devices, the length of traveling path balancing the load on a specific path and the reliability of this path will also greatly affect the lifetime of a WSN. Data in a WSN travels a route from the source node to a selected successor node. It repeats this movement based on specific selection methods until reaching the sink node. Routing in a WSN can be classified on network structure or protocol operation. Below are a few explanations for the routing protocols.

4.2.1   Routing Protocols Based on Network Structure

In this section we survey the state-of-the-art routing protocols for WSNs. In general, routing can be classified as flat-based routing where every node plays the same role in the network, hierarchical-based routing where the nodes play different roles, and location-based routing where data are routed according to node positions depending on the network structure [15].

Flat-based routing: Equal roles and funtionalities are assigned to all nodes in a network but the nodes do not follow a structural distribution. They communicate directly and indirectly with a base station that handles communications for large numbers of nodes. A node senses its nearest neighbor nodes and transfers data based on a data-centric routing mode. The main advantages of flat routing are simplicity (no overhead) and scalability based on interchangeability (nodes have the same roles and functionalities).

The network lifetime is maximized by multi-hop routing that balances the loads by restricting the power level at which sensor nodes communicate. However, the main disadvantage of this routing protocol is that fairness among nodes is not guaranteed and this may lead to hotspots when nodes are uniformly distributed. If there is only one sink node in WSN, the energy of the nodes that surround the sink will be consumed and this may shorten the network lifetime. Flooding, gossiping, sensor protocols for information via negotiation (SPIN) and directed diffusion (DD) are flat routing protocols. In flooding-based protocols, a node floods its information in the network. Therefore, all the nodes in the network receive the flooded information and resend it again to their neighbors. Gossiping is based on flooding but only one will randomly be selected to forward the data and this will greatly reduce the consumed energy. SPIN and directed diffusion are mainly eliminating the redundant data and use data negotiation to conserve WSN power.

Hierarchical-based routing: In this routing protocol, different roles, capabilities, and functionalities are assigned to all nodes. All nodes are playing different roles on the network. The network is divided into a number of clusters. Each cluster consists of a number of nodes and nominates only one node to be the head of the cluster. Messages on the network are sent from nodes only to the cluster head and hence the cluster head forwards this message to the sink. Cluster heads are responsible for managing, collecting, aggregating, and retransmitting data from cluster nodes to the base station. Cluster preservation depends on intelligence of detection and recovery determining the survival of the cluster head.

The cluster head is selected and assigned periodically in order to eliminate the overall consumed energy of the network and to prolong the network life time. The tackled problem of this protocol is massive power consumption in the cluster head. Therefore it is always recommended to periodically rotate the cluster heads among the nodes to ensure uniform energy consumption and to prevent energy hotspots. Energy conservation in clustering greatly contributes to overall system scalability and lifetime and energy efficiency.

The main advantage of hierarchical-based routing is the data aggregation as the data of the node in the cluster can be combined with the cluster head and this will reduce the data redundancy. However hierarchical-based routing has many disadvantages such as being a hotspot as a result of cluster head election, the need of excessive energy of the cluster head, the deployment complexity in order to balance power consumption, and the lack of scalability that increases the messages' overhead in clusters. Some examples of hierarchical-based routing are low-energy adaptive clustering hierarchy (LEACH), threshold sensitive energy efficient sensor network protocol (TEEN), adaptive threshold sensitive energy efficient sensor network protocol (APTEEN), power efficient gathering in sensor information systems (PEGASIS), and minimum energy consumption network (MECN).

LEACH forms clusters of the sensor nodes based on the received signal strength and uses the local cluster heads as gateway to the base station. TEEN is a hierarchical protocol designed to be responsive to sudden and drastic changes in the sensed attributes such as temperature, pressure, rainfall, etc. APTEEN aims at capturing periodic data collections and reacting to time-critical events. PEGASIS forms chains of sensor nodes so that each node transmits and receives, and only one node is selected from that chain to transmit to the base station rather than forming multiple clusters. MECN finds a sub network of the WSN with fewer nodes and finds the minimum global energy required for data transfer.

Location-based routing: The positions of sensor nodes' route data in the network. All nodes in the network are addressed by their locations. Location information used to investigate, select and maintain the optimal route to forward the data packets. It is based on the frequent calculations of distances between nodes and the estimation of consumed energy level. This depends on the frequent updates of the nodes' location information. Power management approaches are used to reduce energy consumption and prolong network lifetime by setting some nodes into sleep mode in its deactivation status.

Routing protocols mainly depend on area partitioning schemes and location information. The advantage of using location information-based routing is the ease and optimization to manage the network as well as reducing the control overhead of the network. However, the main disadvantage is the complexity in designing the sleep mode of nodes. Examples of location-based routing are geographical and energy-aware routing (GEAR) and greedy perimeter stateless routing (GPSR). GEAR mainly focuses on optimizing power consumption by using energy awareness, and neighbor location information to set the forwarding route. GPSR uses the perimeter of the planar graph to find the optimal route for sending the packets. This requires a location service to map locations and node identifiers.

4.2.2   Routing Protocols Based on Protocol Operation

These routing protocols are based on the protocol operation. Different routing functionality can be applied according to the variation of the approach used in the protocol. The taxonomies of different protocol operations-based routing are discussed below.

Negotiation-based routing: This is based on exchanging a number of negotiation messages between interconnected nodes. The advantage is that it works to reduce data redundancy and prevent information duplication. An example of negotiation-based routing is sensor protocols for information via negotiation (SPIN) that uses negotiations to address all problems of flooding by utilizing meta-data to succinctly and completely describe sensor data. Sequential assignment routing (SAR) creates multiple trees, each rooted at a one-hop of the sink, to establish multiple paths from each node to the sink. Directed diffusion (DD) uses flooding-based query mechanisms for continuous and aggregate queries.

Multipath-based routing: This protocol is based on finding better paths between sources and sinks to increase routing efficiency and reduce power consumption. The advantage of this protocol lies on reserving the level of consumed power prolongs the network lifetime. It also helps with fault tolerance and quick recovery from broken routes. The network performance will efficiently increase by reducing the transmission delay and reliability will be guaranteed due to lower overheads. However, the disadvantage is the great amount of overhead and energy consumption as a result of sending periodic messages to keep the network paths alive. Establishing and maintaining all trees is expensive. An example of this protocol is multi path and multi SPEED (MMSPEED) that provides QoS differentiation in terms of timeliness and reliability, while minimizing protocol overhead. SPIN uses negotiations to address all problems of flooding as it uses meta-data to succinctly and completely describe sensor data.

Query-based routing: This routing protocol is based on a series of propagated queries between the sources and sink node that sense the traveling paths. The destination node sends the query of interest from a node through the network and the node with this interest matches the query and reports back to the node which initiated the query. The query normally uses high-level languages. The best route is discovered and constructed by the updated information related to each route in the network nodes. The advantage of this routing protocol is that it eliminates redundancy and reduces number of transmissions across the network. The disadvantage is that this protocol may not be the best solution for networks with a need for continuous data transfers such as environmental monitoring. An example of query-based routing is the SPIN that uses negotiations to address all problems of flooding utilizing meta-data to succinctly and completely describe sensor data. DD uses flooding-based query mechanisms for continuous and aggregate queries.

QoS-based routing: This routing protocol is based on balancing all the network constraints to satisfy all the QoS metrics such as energy consumption, data quality, delay, priority level, and bandwidth. The disadvantage is the delay to meet the QoS metrics that consumes massive network energy. It is expensive to establish and maintain all trees on the network. SAR creates multiple trees, each rooted at a one-hop neighbor of the sink to establish multiple paths from each node to the sink. This will minimize the average weighted QoS metric over the lifetime of the network. MMSPEED provides

QoS differentiation in terms of timeliness and reliability, while minimizing protocol overhead.

Coherent-based routing: This routing protocol is based on coherent and non-coherent processing techniques. Energy efficient routes will be selected based on the amount of processing. Coherent routing forwards data after minimum processing to reduce the consumed energy. Non-coherent routing sends data after local processing in each node. The advantage of coherent data processing routing is the energy efficiency due to pre-processing of data and data aggregation. Non-coherent processing involves target detection as data collection and pre-processing take place. Neighboring nodes must be aware of the local network topology. Finally, select central nodes refine information processing. The disadvantage is that central nodes must have enough energy resources and computation abilities.

4.3   Challenges

WSN design poses inherent challenges such as energy available for nodes during operation, maintaining QoS, network lifetime, time delays, complexities of algorithms that find the shortest paths, selecting cluster heads, ease of scalability, ease of introducing a new protocol, and adaptanility to new demands and environments. These challenges are explained below.

4.3.1   Routing Efficiency Limitations

Routing strategies are designed to consider energy constraints of a network. Challenges include the mobility of sensor nodes, types of sensor data, and other control parameters. The mobility of sensor nodes creates impacts such as the Doppler effect and multipath fading. Network design must be intelligent enough to adapt routing protocols to channel requirements.

Dynamic routing is easy in small networks. However, designing routing protocols for networks of huge numbers of nodes is very difficult. Whether transmitted messages are unicast or multicast is another factor in designing routing protocols. Controlling overhead is a difficult process because of the complexity required to manage many parameters.

4.3.2   Energy Limitations

The design of a sensor node depends upon the inherent energy limitations in WSN design, particularly the PHY, MAC, and other layers. During the creation of an infrastructure, setting up the routes is greatly influenced by energy considerations. Since the transmission power of a wireless radio is proportional to distance squared or greater in the presence of obstacles, multi-hop routing will consume less energy than direct communication. However, multi-hop routing introduces significant overhead for topology management and medium access control. Direct routing would perform well enough if all the nodes were close to the sink. Most sensors are scattered randomly over an area of interest and multi-hop routing becomes unavoidable.

4.3.3   Other Factors

Other considerations such as deployment problems and strategies for data aggregation and fusion also influence design. Since the algorithms for deployment and data aggregation are time consuming, it is important to choose the node to implement the algorithms (generally on a cluster head). Factors like cluster head selection determine the overall efficiency of the network design.

4.4   Soft Computing Based Algorithms

4.4.1   Case for Soft Computing Algorithms in WSNs

All the challenges listed can be attributed to one fact. WSNs face significant challenges in the form of energy resource constraints due to the inbuilt design framework which does not include a centralized power supply. All the layers of a WSN aim at lowering energy consumption. Energy consumption and power management approaches were recently addressed to tackle this problem. Optimal routing and energy optimization significantly affect WSN performance and guarantee the extension of the network lifetime.

Huge numbers of nodes are required in the design of a WSN. The amount of control data the nodes exchange to control a network is comparable to the amount of sensor data required; this is a fairly new finding. Control data accounts for significant energy consumption that WSNs cannot affort. As WSNs scale up to tens of thousands of nodes, network designers face design challenges, particularly in managing energy constraints.

Soft computing comes to the rescue. Soft computing approaches are very useful in designs of intercoupled, very scalable, parallelized systems. The next section covers soft computing paradigms from a WSN view. Soft computing will revolutionize WSN designs.

Sensor energy use and other WSN constraints require smart routing to balance energy consumption among nodes, thus prolonging network life and ensuring coverage. Smart techniques enhance the effectiveness of WSNs [20]. Various soft computing paradigms for optimizing WSN routing by reducing power consumption and designing other innovations are being explored.

4.4.2   Different Soft Computing Algorithms

The soft computing paradigms such as reinforcement learning (RL), swarm intelligence (SI), evolutionary algorithms (EAs), fuzzy logic (FL), and neural networks (NNs) have been applied to different WSN applications and deployment based on their different characteristics.

Figure 4.1 shows comparison of the different soft computing algorithms for WSNs.

4.4.2.1   Reinforcement Learning (RL)

Reinforcement learning (RL) approach uses feedback to teach behavior and adapt the system parameters to maximize the system performance. This technique is used widely in many research areas. Many RL algorithms are related to dynamic programming techniques, differing from classical techniques in that RL algorithms do not assume the Markov decision process. Position of sensor nodes in the network and the distance from sink node greatly affect the level of energy consumption in each node. Each node needs an adaptive energy consumption and optimization algorithm to improve performance and preserve the energy of the network. The idea behind the usability of RL lies in the Q-learning algorithm that is able to construct better solutions for distributing and dynamic problems such as clustering and routing with minimal communication and computational requirements. There are three variations of Q-learning approaches in the literature aiming at different parameters of WSNs.

1. Delay: The original Q-learning approach [1] proposed in 1994 is relevant to many problems in WSNs. It offers an efficient way of achieving dynamic load balancing by understanding the transport delay by the nodes. It helps articulate the architecture that leads to the maximum utilization of resources. One of the main criticisms of the original Q-routing algorithm concerned its exploration behavior in certain situations. A novel varying Q-routing algorithm has been proposed [2] to overcome this limitation.

2. Energy expenditure: In this variation, an RL algorithm is proposed for balancing energy expenditure in wireless sensor networks. The approach is based on using a node's energy level as a metric, an idea first proposed in the context of WSNs. These algorithms [3,4] use a multi-sink scenario and each node calculates the cost of complete routes that service all sinks.

3. Reducing control payloads: Q-routing with compression [5] attempts to aggregate messages as early as possible in the routing process and tries to compress them before sending them to a single sink. Q-learning techniques are used to find the best compression path toward the sink. The algorithm is fully distributed and its concepts can be applied to the field

Property	Reinforcement Learning	Swarm Intelligence	Evolutionary Algorithms	Fuzzy Logic	Neural Networks
Applicability	High	High	Low	Medium	High
Complexity	Less	Less	More	More	More
WSN benefits	Optimal path and load balancing	Optimal path	Computing cost link functions	Cluster head detection	Energy balancing
Scalability	Easy	Difficult	Easy	Easy	Hard
References	1-5	6,7	11,12	11,12	13,14

FIGURE 4.1: Comparison of soft computing algorithms

of WSNs. While this work has not been used in the experiments presented in this chapter, applying some of the techniques suggested in this chapter should be considered for future research.

4.4.2.2 Swarm Intelligence (SI)

Swarm intelligence (SI) is the technique in which behavioral patterns of large groups are analyzed and applied in the design of future systems. Ant colony optimization is one example of swarm intelligence. After the traditional Q routing methods, SI is the second most powerful soft computing paradigm very much suited for WSN routing. SI addresses the management of collective behaviors of highly dynamic and distributed elements in decentralized and self-deployed systems. The central idea is to optimize the design patterns of large scale WSNs by finding the shortest paths between source and destination in existing standard routing algorithms. Researchers invented a number of techniques such as ant-based routing protocols [7] and slime mold [6] inspired models. In [6], two mechanisms of the slime mold tubular network formation process were mathematically modeled and applied to the routing scenarios in WSNs. A slime mold is a protozoan which distributes nutrients throughout its body intelligently along the tubes of varying radii without centralized control. This decentralization approach differs SI from RL. The reachability in the large scale networks and the unreachable islands in the network pose significant challenges to this approach.

The ant colony works on a dynamic system, especially in a changing graph topologies environment. Examples of such systems include computer networks and artificial intelligence simulations of workers. In [7], the detailed performance analysis of standard ant-based protocols applied to WSNs is presented. The modeling of WSNs as an ant colony opens up a huge scope of research in this area. The parameter under consideration makes a significant impact.

More efficiency can be extracted by combining the techniques intelligently. For instance, where the WSNs are slightly dynamic but very distant from each other, using the slime mold approach for the entire network but ant-based approaches in the island region would result in greater optimization of design process.

4.4.2.3 Evolutionary Algorithms (EAs)

This soft computing paradigm is an artificial intelligence computational optimization algorithm for population heuristics. Genetic algorithm (GA) is one of the most popular EA algorithms that mimic natural evolution. Although best suited for the route optimization, the concept of genetic algorithms, applied for highly dynamic WSN environments does not lead to fruitful results in comparison to RL and SI techniques since the genes of the genetic operator are mostly from the individual while the quality level of the individual determines the efficiency of the algorithm.

The evolutionary algorithms as applied to WSN routing are active research areas. In [8], it is shown that the algorithm balances energy consumption and extends the network life cycle; the network efficiency of WSNs is improved with EAs. In [9], a differential evolution-based routing algorithm is designed for environmental monitoring wireless sensor networks. In [10], genetic algorithm is used to optimize the minimum cost function for calculating the next best hop to optimize the energy wastage.

4.4.2.4   Fuzzy Logic (FL)

This soft computing paradigm is a flexible mathematical and computational model that deals with fuzziness and uncertainty of data. A WSN is an uncertain environment with some insufficient data and it needs special decision making. It needs flexible and tunable procedures that deploy routing and enhance network lifetime. Using fuzzy logic will avoid complex mathematical modeling and provide WSNs with great flexibility to deal with uncertainty and imprecision. The critical tradeoffs in WSN include minimized consumed energy versus transmission route, multi-hop versus direct communication, and computation versus communication.

In [11], parameters such as closeness of node to the shortest path, closeness of node to the sink, and degree of energy balance are put into fuzzy logic systems. Appropriate cluster head node election can drastically reduce energy consumption and enhance the lifetime of a network. In [12], a fuzzy logic approach to cluster head election is based on energy, concentration, and centrality.

4.4.2.5   Neural Networks (NNs)

This soft computing paradigm is a learnable arithmetical algorithm that maps a complex relation between input and output based on supervised learning methods in different environments. Applying the neural network paradigm in context of wireless sensor network provides understanding of WSN. The architectural match between neurons in ANNs and sensor nodes in WSNs as well as the connectivity paradigm presents a great analogy between WSNs and ANNs. Replacing traditional signal processing algorithms in WSNs by simple computation ANN carries out efficient implementation as well as resource reduction in WSNs.

ANN strongly proved its compatibility in WSNs and effectively ensures its prolonged existence. It is able to predict and reduce consumed energy in sensor nodes and the energy for each route. While clustering is significant to improve the network lifetime, we need to reduce the consumed energy and increase the scalability of the sensor network. Pre-assignment and election of cluster heads can produce scalable sensor networks while reducing the consumed energy can efficiently achieve cluster-based routing. Sensing nodes can be developed using ANN and reduce energy through power measurements and prediction. With the simplicity of NNs, we cannot strongly agree its compatibility in WSN routing. NNs have some challenges that can't cope with the nature and the frequent changes of WSN properties. It requires an offline learning phase and too many calculations that can't easily adopt with the WSN topology.

In [14], different NN techniques have been employed to achieve the energy efficiency. The results were quite competitive to the other soft computing techniques discussed, and [13] is one more application of NNs in WSN.

4.5   Research Directions

Since soft computing is an emerging technology in the field of wireless sensor networks, we suggested the possible research directions for students to explore further.

1. IoT: When the evolution of wireless sensor networks to the Internet of Things (IoT) is taking place, there are a whole new set of challenges that come into play. Soft computing techniques should be fine tuned to meet the challenges.

2. WSDN: This is an emerging area that will facilitate the implementation of soft computing based routing strategies and open a plethora of research opportunities.

3. Cloud computing and big data: As the number of nodes increases in WSNs, the data exchanged will reach tera-bytes each day. In such a situation, the role of soft computing will be critical.

4.6   Conclusion

This chapter summarized the importance of soft computing in routing protocols of wireless sensor networks.

Bibliography

[1] J. A. Boyan and M. L. Littman. Packet routing in dynamically changing networks: A reinforcement learning approach. In *Advances in Neural Information Processing Systems*, vol. 6, pages 671–678. Morgan Kaufmann, 1994.

[2] M. Fakir, K. Steenhaut, and A. Nowe. Qos based routing in atm networks using a variant q-routing with load balancing. In *Proceedings of the Fifth Multi-Conference on Systems, Cybernetics and Informatics*, 2001.

[3] A. Forster and A. L. Murphy. Balancing energy expenditure in wsns through reinforcement learning: A study. In *Proceedings of the 1st International Workshop on Energy in Wireless Sensor Networks WEWSN*, 7 pages, Citeseer, 2008.

[4] A. Forster. Machine learning techniques applied to wireless ad-hoc networks: Guide and survey. In *Proc. of the IEEE International Conference on Intelligent Sensors, Sensor Networks and Information*, 2007.

[5] P. Beyens et al. Routing with Compression in Wireless Sensor Networks: a Q-learning Appoach. In *Proceedings of the 5th European Workshop on Adaptive Agents and Multi-Agent Systems (AAMAS)*, 2005.

[6] K. Li et al. Slime mold inspired routing protocols for wireless sensor networks. *Swarm Intelligence*, 5.3-4 (2011): 183-223.

[7] A. M. Zungeru, A. Li-Minn, and K. P. Seng. Performance evaluation of ant-based routing protocols for wireless sensor networks. arXiv preprint arXiv:1206.5938 (2012).

[8] L. Guo and Q. Tang. An improved routing protocol in WSN with hybrid genetic algorithm, *IEEE International Conference on Networks Security Wireless Communications and Trusted Computing (NSWCTC)*, Vol. 2, 2010.

[9] X. Li, L. Xu, H. Wang, J. Song, and S. X. Yang. A Differential Evolution-Based Routing Algorithm for Environmental Monitoring Wireless Sensor Networks, *Sensors*, 10, No. 6, pp. 5425–5442, 2010.

[10] S. Nesa, M. L. Valarmathi, and T. C. Neyandar. Optimizing Energy in WSN Using Evolutionary Algorithm, International Conference on VLSI, Communication and Instrumentation (ICVCI), *International Journal of Computer Applications (IJCA)*, No. 12, 2011.

[11] H. Jiang, Y. Sun, R. Sun, and H. Xu. Fuzzy-Logic-Based Energy Optimized Routing for Wireless Sensor Networks, *International Journal of Distributed Sensor Networks*, vol. 2013, Article ID 216561, 8 pages, 2013. doi:10.1155/2013/216561.

[12] I. Gupta, D. Riordan, and S. Sampalli. Cluster-head election using fuzzy logic for wireless sensor networks. In *Proceedings of the 3rd Annual Conference Communication Networks and Services Research*, 2005.

[13] A. Abdalkarim, T. Frunzke, and F. Dressler. Adaptive distance estimation and localization in WSN using RSSI measures. *Euromicro Conference on Digital System Design Architectures, Methods and Tools*, 2007.

[14] N. Enami et al. Neural network based energy efficiency in wireless sensor networks: A survey. *International Journal of Computer Science and Engineering Survey*, 1.1 (2010): 39-53.

[15] M. Sharawi et al. Routing Wireless Sensor Networks based on Soft Computing Paradigms: Survey. *International Journal on Soft Computing, Artificial Intelligence and Applications (IJSCAI)*, 2.4 (2013).

[16] K. Akkaya and M. Younis. A survey on routing protocols for wireless sensor networks. *Ad hoc Networks*, 3.3 (2005): 325-349.

Chapter 5

Fault-Tolerant Routing in WSNs

P. Venkata Krishna

Sri Padmavati Mahila University

V. Saritha

VIT University

5.1   Introduction

A wireless sensor network (WSN) is a network which is ad hoc and self-configured. The nodes in WSNs may be prone to faulty performance for many reasons. The major concern is how to withstand the fault-prone nodes at the protocol design level. Faults significantly degrade the performance of WSNs and affect network survival time with unnecessary energy consumption and overhead. To make a network robust and reliable, fault tolerance has become an immediate need considering energy consumption is an important factor. One of the possible ways of making a network fault tolerant is routing using alternate paths. In this chapter, issues of fault-tolerant routing are discussed with a focus on soft computing techniques.

Devices equipped with sensors form a WSN by connecting sensors with each other without wires. These devices are usually cheap and also have battery life. These devices are referred to as sensor nodes in WSNs.

WSNs can be formed at any location (ground, air, water). Their sensors seize the required data and send it to the base station or server node. Power resources in sensor nodes cannot be replaced as easily or frequently as they can in cellular or ad hoc networks that presents no risks from energy limitations. Limited power is one of the main characteristics of WSN nodes that must be considered in designing appropriate protocols.

Advances in scaling down the dimensions of cost-effective electronic devices have helped meet the needs for smaller devices that can handle substantial amounts of data and mine valuable data in different situations. However, designing for small dimensions limits the designs and programming capacities of sensor nodes and thus affects their storage capacities, processing abilities, and power needs. Limited power supply is one of the most important challenges.

WSNs must work independently for durations extending from days to years [34]. They are more vulnerable to recurrent and unpredictable faults than other networks. Designers face the tradeoff of extending system lifetime by limiting the power of individual nodes and maintaining reliability by installing intricate fault-tolerant procedures.

Other dynamics such as antenna position, obstacles, signal strength, interference, and atmospheric conditions influence communications among sensor nodes. Sensor nodes are also susceptible to failures of links, hardware, software, and power supplies. Sensor systems must tolerate tough natural conditions like poor weather and fire. They are called malicious nodes when they exhibit misbehaviors or simply crash. For these reasons, building fault tolerance into sensor nodes requires additional considerations [34].

Fault tolerance is a procedure for handling situations like breakdowns and is vital for systems that have few maintenance options. Fault tolerance empowers a network to continue working appropriately despite a fault or part breakdown.

WSNs require thoughtful designs because of their important uses in military, ecological, social, engineering, and other vital areas. WSNs can be considered unique sets of wireless ad hoc networks. A fault-tolerant system must perform adequately even when faults such as node failures occur [5]. Widespread research continues in this area.

Adaptation to non-critical failure at different stages has been studied [10]. Failures can occur in any one of four levels: hardware, network coomunication, software, and application [14]. A failure of a CPU, battery, memory, interface, or detecting component is a fault at the hardware level. The two kinds of software in a sensor node are system software (operating system) and middleware (communications, accumulation, and routing). Any problem with these components is considered a system software level fault. Link damage is a network layer fault.

Environment and radio interface are the main causes of damage to links in SNs, assuming no faults occurred at hardware level. Aggressive techniques for rectifying faults and retransmitting data are essential for maintaining communication quality in WSNs, but correcting faults may cause delay. The designer must consider a tradeoff between effectiveness and adaptions to non-critical failures.

Finding another path in the event of a failure in an existing path is considered an application layer fault tolerance mechanism. the mechanism is not adaptable to diverse situations so fault tolerance on the application layer must be handled individually for each application.

WSNs typically utilize hundreds of unmanned battery-powered sensor nodes arranged according to plan or arbitrarily to perform certain tasks or gather data from their environment [3] [21]. WSNs attract researchers from interdisciplinary fields because of overwhelming demands from industry, government, and academia for carrying out their tasks with precision in hostile environments. The result has been the extensive deployment of small low-cost sensor nodes in large networks installed to monitor events and gather data in remote areas. The data is transmitted to a base station for analysis and to enable decision making. WSNs are suitable for many applications because the number of sensors they contain can work together to sense large areas and collect information about events of interest. WSNs have improved our ability to monitor conditions on difficult and complex terrains.

Sensors are self-monitored; they have limited storage space and power for processing data from other nodes. They consist of miniscule, inexpensive complementary metal oxide semiconductor (CMOS) components. CMOS devices replace the older microelectromechanical system (MEMS) devices and one advantage is that CMOSs do not require human intervention.

Because of their low deployment cost and advanced components, WSNs are used in multimedia applications and monitoring of ecological conditions (temperature, sound, vibration, atmospheric pressure, pollutant levels, and animal movements). They can function in hostile environments like battlefields, high altitudes, and deep sea trenches where humans cannot collect data.

WSNs are self-configured ad hoc networks. They are fault-prone at the protocol design stage and this is a point of concern. Faults degrade performance, influence node survival time, increase energy consumption, and computational overhead. Fault tolerance that does not impact energy consumption is an immediate requirement for WSN designs.

5.2   Learning Automata-Based Fault-Tolerant Routing Protocol

5.2.1   Learning Automata

5.2.1.1   Introduction

The learning component of Learning Automata (LA) [35] means attaining information while a mechanism/program (automaton) is utilized and determining activities to be performed based on the knowledge attained. The three modules of the LA model are the automaton, the environment, and the reward, or penalty structure. The automaton is the system by which a process learns from its past. The environment is the place where the automaton operates and the environment reacts progressively or adversely to movements of the automaton. The responses determine whether the automaton is rewarded or penalized.

After the automaton performs learning for a certain period, it will be able to determine which actions are optimal and can be executed on the environment. LAs have been studied extensively [22] [27]. Various applications of LA to networks have also been reported [16] [17] [19] [26].

5.2.1.2   Automaton

The quintuple used to characterize learning automaton is $\{Q, A, B, F, H\}$, where

- $Q\{q_1, q_2, q_3, ..., q_n\}$ is the finite set of internal states where q_n is the state of the automaton at instant n.

- $A\{\alpha_1, \alpha_2, ..., \alpha_n\}$ is a finite set of actions performed by the automaton where n is the action performed by the automaton at instant n.

- $B\{\beta_1, \beta_2, \beta_3, ..., \beta_n\}$ is a finite set of responses from the environment to the automaton where n is the response from the environment at an instant n.

- F is a function that maps the current state and input of the automaton or the response from the environment to the next state of the automaton. $Q \times B \to Q$.

- H is a function that maps the current state and response from the environment to determine the next action to be performed. $Q \times B \to A$.

5.2.1.3 Environment

The environment is the medium in which the automaton operates. Scientifically, an environment can be represented by a triple $\{A, B, C\}$. A and B have already been defined above. $C = \{c_1, c_2, ..., c_r\}$ is a set of penalty probabilities, where element $c_i \in C$ corresponds to an input action α_i.

5.2.2 Cross Layer Design

Wireless systems are broadcast media and thus the information sent by one node will be heard by all the other nodes. This is an important point for protocol designers of both wired and wireless networks to consider.

Layered architectures like the seven-layer open systems interconnect (OSI) model define the hierarchy of services provided by individual layers. This means that each layer delivers certain services that can communicate only with services in neighboring layers. For example, a service is defined in layer 2 can interact only with services in layers 1 and 3 via ports and cannot interact with services in other layers. Layered architectures like OSI are not adaptable to some networks because of broadcast nature of wireless communication channels and stringent real-time requirements [4] [11] [12] [29]. Functionalities of the various layers are strictly coupled due to the shared nature of the wireless communication channel.

A cross layer architecture [4] [11] [12] [29] would be effective for delivering the necessary flexibility and meeting the limitations of wireless networks. Cross layer systems adapt protocols to a wireless context by sharing network status among layers and overall optimization instead of optimization at different layers. This avoids duplication of effort in collecting internal information and results in a more efficient design.

Cross layer design is intended to utilize the interactions among system parameters in various layers to achieve optimal performance for time-varying WSNs. Cross layer transmission of data in WSNs is thus an optimization issue. Cross layer design allows services defined in one layer to communicate with those defined in other layers via specific interfaces. The cross layer design chosen depends on the type of network application and performance gain required. The literature [4] [7] [11] [12] [29] [31] shows many examples of current work in cross layer design.

To facilitate cross layer interaction [11], several approaches such as design of new interface among layers and use of shared data structures among layers can be used. Cross layer design is effective for networks, particularly for communication architecture. Performance gains achieved by cross layer design should exceed the cost of modular system architecture but a protocol designer should not overlook the costs of cross layer designs.

5.2.3 Dynamic Sleep Scheduled Cross Layered Learning Automata-Based Approach to Prevent DoS Attacks

This section discusses a dynamic sleep scheduled cross layered learning-based fault-tolerant routing (DCLFR) protocol for WSNs. The design helps route data efficiently across a network even if faults occur by maintaining multiple paths between pairs of source and sink nodes based on certain parameters. Learning automata (LA) in the system minimize network overhead through multipath routing and cross layer design that saves energy by controlling the node sleep schedules. This section also describes an approach for minimizing overhead and enhancing energy efficiency.

5.2.3.1 Network Model

A wireless network consists of a source, sink, and intermediate sensor nodes. The sink node is assumed to have unlimited energy; the energy levels of other sensor nodes vary. The many-to-one network model is based on sending data gathered from multiple sources to the sink.

An ad hoc sensor network is represented using a graph $W = (V, E)$, where V is the set of vertices and E the set of edges. Vertices of the graph represent nodes and edges represent the links between the nodes. A path is a set of vertices connected to each other from a vertex (which can also be a source) to a destination (sink). Faults may arise arbitrarily in any node in the network. All the edges in the graph are assumed to be bidirectional, i.e., if $(v_i, v_{i+1}) \rightarrow e_i$, then $(v_{i+1}, v_i) \rightarrow e_i$ also exists where $v_i, v_{i+1} \in V$ and $e_i \in E$. There are two modules called routing module and LA module in each node $v_i \in V$. A table with LA information in the routing module is updated and shared among neighboring nodes [18]. The LA module in each node is autonomously operated. The communication between two neighboring nodes in the network is shown in Figure 5.1.

The cross layer architecture presented involves the physical layer, MAC layer, and network layer. The routing and scheduling information is shared by the network and MAC layers and the routing and energy level details are shared between network and physical layers of the node. DCLFR is independent of the network topology.

5.2.3.2 Learning Automaton Model

An automaton is placed at each node in the sensor network. The learning automata model used here is taken from Misra et al. [18]. Every forwarding node receives an acknowledgment for every successful packet delivery and updating is applied to the path over which the acknowledgment is sent. An S-model variant of LA is used here.

Need for learning system to design fault-tolerant routing protocol: A proficient fault-tolerant routing algorithm is NP-hard [18] as details re-

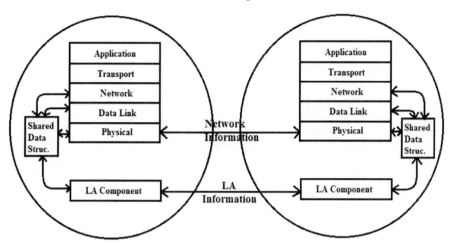

FIGURE 5.1: Interactions of components of node and neighboring node

garding the path are missing. Therefore, it is challenging to design an efficient fault-tolerant routing algorithm for ad hoc networks. Due to the continuously changing nature of these networks, there is a need to minimize the number of iterations needed for the network to choose a path after reorganization and simultaneously decrease the overhead (redundant packets) in the network. This is similar to the 0 or 1 knapsack problem [18]. On one hand, the overhead need is to be reduced so that network resources are not overused with redundant packets, and on the other hand, with a high packet delivery rate a node can be ensured without significantly increasing the overhead in the network. An efficient solution to this problem may be a learning mechanism based on the packet delivery rate, where a node can choose the best path, and which does not introduce large overhead in the network. The mechanism should be scalable and self-learning, so that a large number of nodes can be accommodated.

Need for cross layer design of fault-tolerant routing protocol: Faults in a network [7] [31] degrade the performance of the routing protocols. There is an immediate need for a fault-tolerant routing protocol which can deliver packets at a guaranteed delivery rate even in case of node failure and node reorganization, without significantly affecting the performance of the application running on top of it. Energy limitation [7] [31] is a major constraint for any protocol in wireless sensor networks. Sleep scheduling using cross layer interactions among layers of a network can reduce energy consumption of sensor nodes. For example, a network layer can use physical layer information in making routing decision to route data via paths with greater available energy [11].

5.2.3.3 Algorithm

Generally, a routing algorithm in an ad hoc network begins with path discovery. Obviously, it would be difficult to route data without sufficient information about path availability between the source and destination. The route discovery mechanism of DCLFR is similar to that for AODV, but DCLFR uses a strategy to reduce control packets in the network.

An automaton is stationed on each node to update the goodness value of the node based on a reward or penalty scheme [18]. When the goodness value of a path exceeds the threshold value, the path is considered optimal for transmitting data and the remaining paths are put to sleep. During sleep, nodes switch their transmitters off and only their receivers are active. As a result, their energy is saved and network lifetime is increased.

The path with the highest goodness value is selected for transmitting data; the energy of other paths is reduced to 10% [18]. If no path has goodness value higher than the threshold value, multiple paths are used for data transmission. That increases energy consumption but provides better connectivity.

5.2.3.4 Strategy to Reduce Control Packet Overhead

The key idea of reducing overhead is that an intermediate node does not forward a route request (RREQ) packet initiated by a particular source for a specific destination more than once. This will ultimately reduce the overhead and thus make a network more energy efficient. The new path will be determined using the route discovery mechanism when there is no path between source and destination. A route discovery mechanism similar to AODV is employed for path discovery but route discovery intermediate nodes do not forward route request (RREQ) packets destined to a particular node from the same source more than once.

During the route discovery a source node will send route request packets to all its neighbors if no path exists between source and destination. Each intermediate node then sends route request packets to all its neighbors. The process is repeated until the RREQ packet does not reach the destination. Intermediate nodes in ad hoc networks send multiple route requests via different paths from source to destination. In DCLFR, such packets are forwarded only once; later packets coming from neighbor nodes are discarded.

A packet is forwarded as soon as it emits from one of the neighboring nodes. Its sequence number is stored on a list maintained at the node. Intermediate nodes maintain lists of all discarded packets.

For example, in Figure 5.2 if the route request packet from source A for destination H via route A to B to D reaches to E before route A to C then only that packet will be forwarded. Making an entry in the list at E will silently drop the route request packet coming from A to C. But, whenever a route reply comes from H to E, E will send a reply via both D to B to A and C to A by checking the list of discarded RREQ packets at E. In this way, DCLFR prevents flooding of a network by control packets, which reduces energy con-

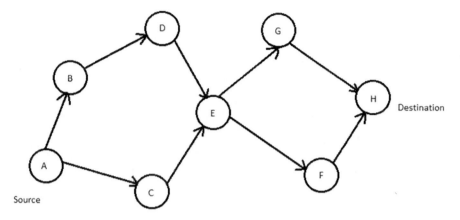

FIGURE 5.2: System strategy for control packet overhead

sumption of the network. However, during selection of packets coming from such a path, the selected path may not be the shortest path (as in above example). Hence, while forwarding packets from source to destination, the least congested path is selected. Therefore, such a scheme will reduce end-to-end delays in the network.

5.2.3.5 Learning Phase

During the learning phase, automata stationed at nodes update the goodness values of nodes and paths in the network. A node is penalized if the packet is not delivered successfully; otherwise it is rewarded. The data structures of LAFTRA [18] such as goodness value table, goodness update message, reward and penalty structure are used in implementing the learning phase of DCLFR. The LA component maintains a goodness table at each node that contains goodness value of all paths from the node to the destination node [18]. The table consists of the entries such as node identification, next hop, update sequence number, and path goodness. The entry node shows the destination node. Next hop indicates the neighbour node, which is part of the path with highest goodness value to the destination node from the entry node. Update sequence number is used to track the updates in the table. The table is updated only if an update message with higher update sequence number arrives. Path goodness denotes the goodness value of the best path from the current node to the destination calculated on the basis of the reward and penalty scheme [18]. The update message sent by the neighboring nodes about the goodness value of the path should be short, in order to minimize the network overhead, and should be sent on a regular basis to avoid redundant entries.

When a packet is transmitted successfully the reward scheme of LA is invoked at every sensor node [18]. The main difference in the reward scheme of DCLFR when compared to LAFTRA [18] is that the energy factor of the node is taken into account while calculating reward parameter for a node.

Algorithm 1 Reward function

$G = G + R \times E$

If current node $=$ destination

   $Y = G$

else if $Y <= T$

   $Y = \eta \times G + (1 - \eta) \times Y_{n-1}$

Algorithm 2 Penalty function

$G = G \times P \times E$

If current node $=$ destination

   $Y = G$

else if $Y <= T$

   $Y = \eta \times G + (1 - \eta) \times Y_{n-1}$

In the reward scheme given above, G is the goodness value of the node, R is the reward constant, E is the energy factor, Y is the goodness value of the path, T is the threshold goodness value of the path, and η is a constant which is introduced to estimate the weightage of goodness value of current node in selected path. The reward and penalty functions are shown above.

Higher value of η will give more weight to a current node in the goodness value of the path. The energy of the node affects the rewarding scheme. If the energy of node is low, the reward will be less and the goodness value of the node will increase slowly. If there is a path with high goodness value but nodes in the path have low energy, such a path will be given lesser preference as compared to a path consisting of nodes with high energy because there are more chances of faults where energy is very low as nodes may run out of power. The multiple paths will converge into single path with a greater speed if the value of R is high. The single path obtained with fast convergence is not sufficiently robust whereas the path obtained with slow convergence because of lower value of R is more robust.

The reason for the path with slow convergence to be more robust is because this path is determined after many cycles of learning which obviously makes it better. Ideally, R should be around 0.1 which will ensure a robust and optimal path selection without much lowering of convergence rate.

If the packet is not transmitted successfully, the node will be penalized by the LA module. At the same time, the complete path, i.e., all the nodes in the path are penalized. Sensor nodes make up the environment. LA at each node rewards or penalizes the node based on packet delivery. If P is chosen to have too low value, the path which is penalized is selected even after a fault in the network. The packet failures which happen unexpectedly and without any fault in the network cannot be tolerated by the network when the P is too high [18].

5.2.3.6 Sleep Scheduling

Cross layer interaction for DCLFR protocol is done by merging MAC and network layers which helps sleep scheduling of nodes in the network [6] [15] [24] [25] [28] [32] [33]. When the goodness value of a particular path crosses the threshold value, that path is used for data transmission and nodes in remaining paths switch to sleep mode. This is done when nodes in alternate paths do not sense any data or control packets addressed to them for some threshold time, TI. If a node does not receive any RTS packets for TI time and the goodness value of path is below threshold value, the node in the path will switch to sleep mode. However, this can lead to a situation when goodness values of all paths are below the threshold value and multipath transmission is being used for transmitting data between source and destination. During this multipath transmission, if there is no data to send from source to destination, all the paths will go to sleep. This will increase the end-end delay while transmitting the next packet. Unlike S-MAC [32] [33], nodes in a proposed sleep scheduling algorithm undergo sleep and wake up depending on the goodness value of a path (Y), hence reducing collisions by unnecessarily waking up the senor nodes of the path which is inefficient to transfer data to the sink node. When a node wants to transmit a data packet, it first tries to acquire the channel for transmission. This happens in asynchronous mode in the following manner.

Procedure for occupying channel for transmission:

1. Listen to channel transmitting the data.

2. Transmit preamble to occupy the channel.

3. Send the RTS packet.

4. Wait for the CTS packet from receiver.

Once the neighboring nodes receive a request to send (RTS) packet, they send a clear-to-send (CTS) packet to qualify as a receiver and confirm the sender to send data packet. This happens in the following manner.

Procedure for qualifying as receiver:

1. Listen to channel for data transmission.

2. After receiving the RTS packet, send the CTS packet to sender.

However, there may be a case when multiple nodes reply with a CTS packet. This leads to collisions and will reduce the efficiency of the algorithm. DCLFR is tested for (a) sleep mode, where only the sensor nodes present in the optimum path are used for data transmission, and nodes in remaining paths are put to sleep, with sleep scheduling, and (b) nosleep mode, which does not utilize sleep scheduling. The DCLFR's performance results using both modes have been compared with ENFAT-AODV and AODV [8].

5.3   Ant Colony-Based Fault-Tolerance Routing

In 1992 Marco Doringo proposed the first algorithm based on the behavior of ants searching for food [36]. The basic idea was later improved to resolve issues that arose from applications in various fields. This section discusses a routing based on the improved ant colony optimization (ACO) technique and cross layer design. It also explains how the ant colony routing can be made fault tolerant.

The cross layer component of the algorithm places communication functions in the physical, MAC, and the network layers so that the optimal forwarding node can be chosen. The parameters communicated among the layers via cross layer are available energy and time stamp. The dynamic duty cycle is implemented at the MAC layer to reduce congestion and conserve energy. The duty cycle changes based on time needed to transmit [1].

The traditional ACO algorithm chooses the optimal path but also increases network traffic, overhead, and energy consumption as the node information is carried by virtual ants from the source node to the destination. The improved ACO protocol utilizes two ants designated data ant and search ant.

When a node has a packet to transmit, data ant is called to transfer the packets. The data ant verifies the pheromone value in the neighboring table. The best value is selected to determine the next forwarding node. If the data does not have sufficient information to calculate the best pheromone value, the search ant is called to collect the required data from neighbors and update the neighboring table.

To improve this process through fault tolerance, the network must maintain an alternate forwarding node displaying the second highest pheromone value in readiness instead of recalculating the original pheromone value. Maintaining an alternate node does not require extra time; the highest and second highest pheromone values can be determined at the same time.

5.4   Neural Network-Based Fault-Tolerant Routing Algorithms

A neural network (NN) is an expansive framework comprising parallel or distributed processing segments called neurons associated in a graph topology.

The weights are associated with the links among these neurons. The weight vectors are called synapses. The input layer is connected to the output layer using these synapses. In fact, the information of NNs is stored on weights of its associations because they have no memories. Artificial neural networks can characterize the input data indirectly, or the training makes the arithmetic algorithms of neural network study the confounded mappings between input and output. The learning in neural networks takes place with the help of examples in many cases. The set of correct input and output data is fed to the network which learns by these examples and returns correct solutions.

5.4.1   Neural Network Approach for Fault Tolerance

This section describes an approach [13] based on neural networks to attain reliable and fault-tolerant transmission (NNFT-BAM). This approach is also based on bi-directional associative memory (BAM). In the processing of developing procedures which are energy efficient for WSNs, there is benefit from use of neural networks. The neural networks minimize size to make the communication process inexpensive and conserve energy. At the same time, the similarity between WSNs and ANNs is another cause for using neural networks in WSNs [23]. The following are the reasons for which the cooperative ARQ may fail in transmitting packets successfully to the destination:

- Noise interruption even though there are required numbers of forwarding nodes between the source and destination.

- There is no possibility of forwarding the packet further toward the destination.

NNFT-BAM resolves both the above issues. The application of NNFT-BAM is where the next action going to take place is based on a known set of actions. The packet size is chosen appropriately if the packet size is small, more transmissions are required or there is a chance of packet drop. There are two ways in which the reliable packet delivery can be achieved. They are:

- If a packet size is large, it is compressed and transmitted. As the size reduces, the packet can be transmitted easily with less bandwidth and less time and less probability of packet drop.

- The packet is converted into a vector which is small when compared to the size of the original packet. This vector is transmitted and the original packet is obtained by the destination node by the reverse process.

In the second approach used in NNFT-BAM, the associations are encoded using the discrete BAM neural network. BAM comprises two layers. There are x units in layer 1 and y units in layer 2. Every unit of layer 1 is connected to every unit of layer 2 and vice versa to form a fully connected directed network. The link between the unit of layer 1 and unit of layer 2 is associated with a particular weight. The weight matrix or WM is constructed considering the links from layer 1 to layer 2. Then the weight matrix of links from layer 2 to layer 1 can be obtained as a transpose of WM provided that weights of the corresponding links are the same. The BAM architecture is shown in Figure 5.3. The links from layer 1 to layer 2 are shown as solid lines and the reverse links are shown as dotted lines.

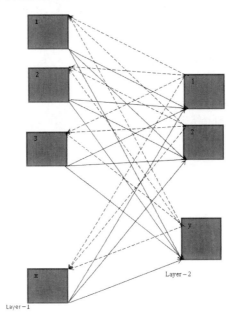

FIGURE 5.3: BAM architecture

The BAM is installed in every source and destination node according to the NNFT-BAM. The associated vector V_i is obtained by passing the packet through the BAM network along with the weight matrix WM before transmitting by the source node. This associated vector is transmitted by the forwarding nodes until the destination receives the vector V_i. At the destination node, the vector V_i is passed through the BAM network again along with the transpose matrix of WM to obtain the packet. BAM is highly fault tolerant of corrupted data because of missing bits rather than mistaken bits.

5.4.1.1   Illustration

Let the original packets be

$$A_1 = \begin{bmatrix} 1 & 0 & 1 & 1 & 0 & 0 & 1 & 1 \end{bmatrix}$$

$$A_2 = \begin{bmatrix} 1 & 1 & 1 & 0 & 0 & 1 & 0 & 1 \end{bmatrix}$$

and the associate packet for transmission be

$$B_1 = \begin{bmatrix} 1 & 1 & 1 \end{bmatrix}$$

$$B_2 = \begin{bmatrix} 0 & 1 & 0 \end{bmatrix}$$

The bi-polar versions can be computed as

$$A_1' = \begin{bmatrix} 1 & -1 & 1 & 1 & -1 & -1 & 1 & 1 \end{bmatrix}$$

$$A_2' = \begin{bmatrix} 1 & 1 & 1 & -1 & -1 & 1 & -1 & 1 \end{bmatrix}$$

$$B_1' = \begin{bmatrix} 1 & 1 & 1 \end{bmatrix}$$

$$B_2' = \begin{bmatrix} -1 & 1 & -1 \end{bmatrix}$$

The weight metric W is given as

$$\sum_{i=1}^{k} (A_i')^T B_i'$$

$$W = (A_1')^T B_1' + (A_2')^T B_2'$$

$$(A_1')^T B_1' = \begin{bmatrix} 1 \\ -1 \\ 1 \\ 1 \\ -1 \\ -1 \\ 1 \\ 1 \end{bmatrix} \begin{bmatrix} 1 & 1 & 1 \end{bmatrix}$$

$$= \begin{bmatrix} 1 & 1 & 1 \\ -1 & -1 & -1 \\ 1 & 1 & 1 \\ 1 & 1 & 1 \\ -1 & -1 & -1 \\ -1 & -1 & -1 \\ 1 & 1 & 1 \\ 1 & 1 & 1 \end{bmatrix}$$

$$(A_2')^T B_2' = \begin{bmatrix} 1 \\ 1 \\ 1 \\ -1 \\ -1 \\ 1 \\ -1 \\ 1 \end{bmatrix} \begin{bmatrix} -1 & 1 & -1 \end{bmatrix}$$

$$= \begin{bmatrix} -1 & 1 & -1 \\ -1 & 1 & -1 \\ -1 & 1 & -1 \\ 1 & -1 & 1 \\ 1 & -1 & 1 \\ -1 & 1 & -1 \\ 1 & -1 & 1 \\ -1 & 1 & -1 \end{bmatrix}$$

$$W = (A_1')^T B_1' + (A_2')^T B_2' = \begin{bmatrix} 1 & 1 & 1 \\ -1 & -1 & -1 \\ 1 & 1 & 1 \\ 1 & 1 & 1 \\ -1 & -1 & -1 \\ -1 & -1 & -1 \\ 1 & 1 & 1 \\ 1 & 1 & 1 \end{bmatrix} + \begin{bmatrix} -1 & 1 & -1 \\ -1 & 1 & -1 \\ -1 & 1 & -1 \\ 1 & -1 & 1 \\ 1 & -1 & 1 \\ -1 & 1 & -1 \\ 1 & -1 & 1 \\ -1 & 1 & -1 \end{bmatrix}$$

$$= \begin{bmatrix} 0 & 2 & 0 \\ -2 & 0 & -2 \\ 0 & 2 & 0 \\ 2 & 0 & 2 \\ 0 & -2 & 0 \\ -2 & 0 & -2 \\ 2 & 0 & 2 \\ 0 & 2 & 0 \end{bmatrix}$$

$$NET = A_1'.W = \begin{bmatrix} 1 & -1 & 1 & 1 & -1 & -1 & 1 & 1 \end{bmatrix} \begin{bmatrix} 0 & 2 & 0 \\ -2 & 0 & -2 \\ 0 & 2 & 0 \\ 2 & 0 & 2 \\ 0 & -2 & 0 \\ -2 & 0 & -2 \\ 2 & 0 & 2 \\ 0 & 2 & 0 \end{bmatrix} = \begin{bmatrix} 8 & 8 & 8 \end{bmatrix}$$

$$Output = F[NET] = \begin{cases} 1 & \text{if } NET > 0 \\ 0 & \text{if } NET < 0 \end{cases}$$

$$Output = F[NET] = \begin{bmatrix} 1 & 1 & 1 \end{bmatrix} = B_1$$

At the receiver side,

$$B_1'.W^T = \begin{bmatrix} 1 & 1 & 1 \end{bmatrix} \begin{bmatrix} 0 & -2 & 0 & 2 & 0 & -2 & 2 & 0 \\ 2 & 0 & 2 & 0 & -2 & 0 & 0 & 2 \\ 0 & -2 & 0 & 2 & 0 & -2 & 2 & 0 \end{bmatrix}$$

$$= \begin{bmatrix} 2 & -4 & 2 & 4 & -2 & -4 & 4 & 2 \end{bmatrix}$$

$$Output = \begin{bmatrix} 1 & 0 & 1 & 1 & 0 & 0 & 1 & 1 \end{bmatrix} = A_1$$

Similarly,

$$NET = A_2'.W = \begin{bmatrix} 1 & 1 & 1 & -1 & -1 & 1 & -1 & 1 \end{bmatrix} \begin{bmatrix} 0 & 2 & 0 \\ -2 & 0 & -2 \\ 0 & 2 & 0 \\ 2 & 0 & 2 \\ 0 & -2 & 0 \\ -2 & 0 & -2 \\ 2 & 0 & 2 \\ 0 & 2 & 0 \end{bmatrix} = \begin{bmatrix} -8 & 8 & -8 \end{bmatrix}$$

$$Output = F[NET] = \begin{bmatrix} 0 & 1 & 0 \end{bmatrix} = B_2$$

At the receiver side,

$$\begin{aligned} B_2'.W^T &= \begin{bmatrix} -1 & 1 & -1 \end{bmatrix} \begin{bmatrix} 0 & -2 & 0 & 2 & 0 & -2 & 2 & 0 \\ 2 & 0 & 2 & 0 & -2 & 0 & 0 & 2 \\ 0 & -2 & 0 & 2 & 0 & -2 & 2 & 0 \end{bmatrix} \\ &= \begin{bmatrix} 2 & 4 & 2 & -4 & -2 & 4 & -4 & 2 \end{bmatrix} \end{aligned}$$

$$Output = \begin{bmatrix} 1 & 1 & 1 & 0 & 0 & 1 & 0 & 1 \end{bmatrix} = A_2$$

5.4.2　Convergecast Routing Algorithm

In this section, the convergecast routing algorithm [30] based on neural networks is presented. The Hopfield neural network (HNN) is used to determine the convergecast route and helps in reducing energy consumption and delay and enhances bandwidth efficiency in the network.

HNN is a form of recurrent artificial neural network invented by John Hopfield in 1982. It can be perceived as a fully connected single layer auto-associative network. Hopfield networks work as content-addressable memory systems with binary threshold nodes. Artificial neurons are used in development of Hopfield networks. There are X inputs and each is associated with a particular weight for artificial neurons. The result is retained until the neuron is updated. The weights between nodes are symmetric. The selection of nodes takes place in a random fashion. There are no hidden nodes or hidden layers. Convergecast is the accumulation of all the data gathered from different nodes in the network moving toward the destination node. Convergecast may precede or work in or parallel with the broadcast operation.

The aim of the algorithm presented in this section is to construct a convergecast tree where all convergecast group members connect. There are three steps in convergecast routing based on the neural network.

- The WSN is divided into clusters and a selection of a cluster head occurs in each cluster.

- Reliable paths are determined.

- The convergecast tree is built using HNN and the reliable paths determined in step 2.

5.4.2.1 Selection of Cluster Head

Clustering is based on the Kohonen self-organizing neural network (KNN) using the system parameters and the needs of the application. Kohonen self-organizing maps (SOMs) are types of neural networks. Tuevo Kohonen developed SOM in 1982. He is a professor emeritus of the Academy of Finland. Supervision is not required. SOMs have a capacity for self-learning using unsupervised competitive learning. Weights are mapped to the associated inputs, maps are included in the SOM. The topological associations among the inputs are conserved precisely [8]. KNN is used to form clusters with nodes of similar input patterns. The adjacency matrix rows are the input patterns and are used for training the KNN. Random weights are given to the links.

- Choose an input pattern and feed it as an input to the KNN.

- Identify the winner output pattern.

- Keep changing the weights of the links associated with the winner output by accomplishing a learning step.

- Repeat the process until there are no changes in the weights.

The final winner output is the cluster head.

5.4.2.2 Determination of Reliable Paths

Path is said to be reliable when the reliable nodes are involved in the path. The reliable path is determined from the source node to the destination node via the cluster head node. Directed diffusion [9] is utilized to determine all the possible sets of paths between all the possible pairs of nodes. After determining all the possible paths among all the nodes, the reliable path of length N hops are determined as follows for one destination to all other nodes.

- The destination node transmits an initial packet which may include data such as reliability of a particular node, number of hops etc.

- The forwarding node keeps a copy of this initial packet, adds necessary information, and forwards to its neighbouring nodes.

- Once the source node receives the initial packet, it replies with an investigative packet toward the destination node.

- The possible paths between the source and destination are determined after the destination node receives the investigative packet sent by the source node.

5.4.2.3   Construction of Convergecast Tree

A single path is chosen among all the possible paths between each pair of source and destination nodes. Then a convergecast tree is built using the set of chosen paths to the destination node.

5.4.2.4   Illustration

Consider the WSN depicted in Figure 5.4.

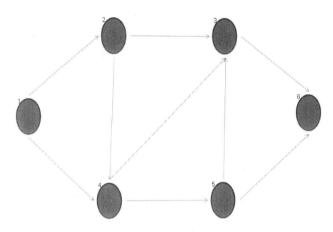

FIGURE 5.4: Sample WSN

Let power in the nodes $\{1, 2, 3, 4, 5, 6\}$ be $\{4, 7, 6, 5, 6, 4\}$ represented as p_i. Let $\{1, 3, 4, 6\}$ be a convergecast group and its size $= 4$. Concergecast group indicates the group of nodes $\{1, 3, 4\}$ are ready to transfer data to destination node 6. The set of possible pairs with node $6 = \{(1, 6)_1, (3, 6)_2, (4, 6)_3\}$.

Let the number of paths of the ith pair be denoted as $NP(i)$. P_{ij} represents the jth possible path for the ith pair and number of hops in a particular path P_{ij} is represented as NH_{ij}.

The possible paths from node 1 to node 6:

P_{11}: $1 \to 2 \to 3 \to 6$ NH_{11}: 3
P_{12}: $1 \to 2 \to 4 \to 5 \to 6$ NH_{12}: 4
P_{13}: $1 \to 2 \to 4 \to 3 \to 6$ NH_{13}: 4
P_{14}: $1 \to 4 \to 5 \to 6$ NH_{14}: 3
P_{15}: $1 \to 4 \to 5 \to 3 \to 6$ NH_{15}: 4
P_{11}: $1 \to 4 \to 3 \to 6$ NH_{16}: 3

The possible paths from node 3 to node 6:
P_{21}: $3 \to 6$ NH_{21}: 1

The possible paths from node 4 to node 6:
P_{31}: $4 \to 5 \to 6$ NH_{31}: 2

P_{32}: $4 \rightarrow 3 \rightarrow 6$ NH_{32}: 2
P_{33}: $4 \rightarrow 5 \rightarrow 3 \rightarrow 6$ NH_{33}: 3

Therefore, $NP(1) = 6$, $NP(2) = 1$, $NP(3) = 3$ with maximum hop count as 4.

The paths P_{14}, P_{21} and P_{31} are selected on the basis of the objective function defined in [30].

Then the convergecast tree constructed is as shown in Figure 5.5.

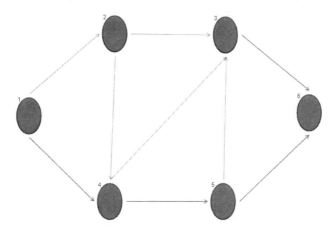

FIGURE 5.5: Convergecast tree for network shown in Figure 5.4

In the procedure to construct a convergecast tree, HNN is utilized.

5.4.2.5 Construction of Convergecast Tree Using HNN

- Determine the cluster head for the nodes in the convergecast group using KNN.

- Identify the path with maximum N hops between every pair of nodes and forward to the cluster node.

- HNN model:

  - Select the starting input to neurons in a random fashion, count = 0.

  - The output of the neuron is obtained using the energy function and sigmoid function of HNN.

  - The above step is repeated until it converges.

  - Using the chosen reliable path, the reliable convergecast tree is built.

Fault tolerance can be incorporated into this algorithm by maintaining the path whose objective function value is second least in the buffer. If the selected path with the minimum objective function value becomes faulty for any reason, the path in the buffer can be used to proceed with the transmission. While the transmission is continuing, an alternate path can be identified and kept in a buffer or the faulty path may be repaired.

5.5   Conclusions

In wireless sensor networks, the sensor nodes may to prone to faults; many challenges surround design of these network protocols. Fault tolerance is a major concern in WSNs and it can indirectly make a system energy efficient. This chapter described four fault-tolerant routing algorithms. Learning automata, cross layer design, ant colony optimization, and other neural network techniques are used. In all the algorithms, reliable paths that present fewer chances of failure are identified and temporary paths are maintained in a buffer. Automatic correction of data is incorporated in the algorithm based on BAM. The study of these algorithms clearly demonstrates that learning procedures and soft computing techniques improve system performance.

Bibliography

[1] M. Abazeed, K. Saleem, S. Zubair, and N. Fisal. CARPM: cross layer ant-based routing protocol for wireless multimedia sensor network. In *Ad-hoc Networks and Wireless*, pages 83–94. Springer, 2014.

[2] N. A. Alrajeh, M. S. Alabed, and M. S. Elwahiby. Secure ant-based routing protocol for wireless sensor network. *International Journal of Distributed Sensor Networks*, 2013.

[3] R. J. Baker. *CMOS: Circuit Design, Layout, and Simulation*, Volume 18. John Wiley & Sons, 2011.

[4] M. Conti, G. Maselli, G. Turi, and S. Giordano. Cross-layering in mobile ad hoc network design. *Computer*, 37(2):48–51, 2004.

[5] M. Demirbas. Scalable Design of Fault-Tolerance for Wireless Sensor Networks. PhD thesis, The Ohio State University, 2004.

[6] O. Diallo, J. J. Rodrigues, and M. Sene. Real-time data management on wireless sensor networks: survey. *Journal of Network and Computer Applications*, 35(3):1013–1021, 2012.

[7] A. J Goldsmith and S. B Wicker. Design challenges for energy-constrained ad hoc wireless networks. *Wireless Communications, IEEE*, 9(4):8–27, 2002.

[8] S. M Guthikonda. Kohonen self-organizing maps. Wittenberg University, 2005.

[9] C. Intanagonwiwat, R. Govindan, D. Estrin, J. Heidemann, and F. Silva. Directed diffusion for wireless sensor networking. *IEEE/ACM Transactions on Networking*, 11(1):2–16, 2003.

[10] F. Koushanfar, M. Potkonjak, and A. Sangiovanni. Fault tolerance techniques for wireless ad hoc sensor networks. Proceedings of IEEE. Vol. 2. IEEE, 2002.

[11] P. Krishna and S. Iyengar. A cross layer based QoS model for wireless and mobile ad hoc networks. *Mobile Communication*, 1:114–120, 2007.

[12] P. V. Krishna, N. Ch. SN. Iyengar, and S. Misra. An efficient hash table-based node identification method for bandwidth reservation in hybrid cellular and ad-hoc networks. *Computer Communications*, 31(4):722–733, 2008.

[13] V. Kumar, R. B. Patel, M. Singh, and R. Vaid. A neural approach for reliable and fault tolerant wireless sensor networks. *IJACSA Editorial*, 2011.

[14] H. Liu, A. Nayak, and I. Stojmenović. Fault-tolerant algorithms/protocols in wireless sensor networks. In *Guide to Wireless Sensor Networks*, pages 261–291. Springer, 2009.

[15] L. DP. Mendes and J. JPC Rodrigues. A survey on cross-layer solutions for wireless sensor networks. *Journal of Network and Computer Applications*, 34(2):523–534, 2011.

[16] S. Misra, P V. Krishna, and K. I. Abraham. Adaptive link-state routing and intrusion detection in wireless mesh networks. *IET information security*, 4(4):374–389, 2010.

[17] S. Misra, P. V. Krishna, and K. I. Abraham. A stochastic learning automata-based solution for intrusion detection in vehicular ad hoc networks. *Security and Communication Networks*, 4(6):666–677, 2011.

[18] S. Misra, P. V. Krishna, A. Bhiwal, A. S. Chawla, B. E Wolfinger, and C. Lee. A learning automata-based fault-tolerant routing algorithm for mobile ad hoc networks. *The Journal of Supercomputing*, 62(1):4–23, 2012.

[19] S. Misra, B J. Oommen, S. Yanamandra, and M. S Obaidat. Random early detection for congestion avoidance in wired networks: a discretized pursuit learning-automata-like solution. *Systems, Man, and Cybernetics, Part B: Cybernetics, IEEE Transactions on*, 40(1):66–76, 2010.

[20] S. Misra, V. Tiwari, and M. S Obaidat. Adaptive learning solution for congestion avoidance in wireless sensor networks. In *Computer Systems and Applications, IEEE/ACS International Conference*, pages 478–484. IEEE, 2009.

[21] S. Misra, I. Woungang, and S. C. Misra. *Guide to Wireless Sensor Networks*. Springer, 2009.

[22] K. S. Narendra and M. A. Thathachar. *Learning Automata: an Introduction*. Courier Corporation, 2012.

[23] F. Oldewurtel and P. Mähönen. Neural wireless sensor networks. In *Systems and Networks Communications, International Conference*, pages 28–28. IEEE, 2006.

[24] Luís M Oliveira and J. JPC Rodrigues. Wireless sensor networks: a survey on environmental monitoring. *Journal of Communications*, 6(2):143–151, 2011.

[25] Luís ML Oliveira, A. F De Sousa, and J. JPC Rodrigues. Routing and mobility approaches in Ipv6 over lowpan mesh networks. *International Journal of Communication Systems*, 24(11):1445–1466, 2011.

[26] B. J. Oommen and S. Misra. A fault-tolerant routing algorithm for mobile ad hoc networks using a stochastic learning-based weak estimation procedure. In *Wireless and Mobile Computing, Networking and Communications, IEEE International Conference*, pages 31–37. IEEE, 2006.

[27] J. Oommen and S. Misra. Cybernetics and learning automata. In *Handbook of Automaton*, pages 221–235. Springer, 2009.

[28] J. J. P. C. Rodrigues and P. A. C. S. Neves. A survey on IP-based wireless sensor network solutions. *International Journal of Communication Systems*, 23(8):963–981, 2010.

[29] V. Srivastava and M. Motani. Cross-layer design: a survey and the road ahead. *Communications Magazine, IEEE*, 43(12):112–119, 2005.

[30] K.N. Veena and B.P. V. Kumar. Convergecast in wireless sensor networks: a neural network approach. In *Internet Multimedia Services Architecture and Application, IEEE 4th International Conference*, pages 1–6, 2010.

[31] Y. Xue and K. Nahrstedt. Providing fault-tolerant ad hoc routing service in adversarial environments. *Wireless Personal Communications*, 29(3-4):367–388, 2004.

[32] W. Ye, J. Heidemann, and D. Estrin. An energy-efficient MAC protocol for wireless sensor networks. In *Twenty-First Annual Joint Conference of the IEEE Computer and Communications Societies Proceedings.* volume 3, pages 1567–1576, 2002.

[33] W. Ye, J. Heidemann, and D. Estrin. Medium access control with co-ordinated adaptive sleeping for wireless sensor networks. *Networking, IEEE/ACM Transactions*, 12(3):493–506, 2004.

[34] M. Yu, H. Mokhtar, and M. Merabti. Fault management in wireless sensor networks. *Wireless Communications, IEEE*, 14(6):13–19, 2007.

[35] S. Senturia, MEMS Reference Shelf (Book), Springer, 2012.

[36] A. Colorni, M. Dorigo, and V. Maniezzo. An Investigation of Some Properties of an Ant Algorithm, *PPSN*, Vol. 92, 1992.

Chapter 6

Mathematical Analysis of WSNs Using Game Theory: A Survey

Hindol Bhattacharya

Jadavpur University

Samiran Chattopadhyay

Jadavpur University

Matangini Chattopadhyay

Jadavpur University

6.1   Introduction

A wireless sensor network (WSN) is a set of sensors spatially distributed to autonomously monitor some environmental conditions. They form a network so as to forward their collected data to a central location [4]. The number of sensor nodes can vary for each network, from a few hundred to several thousands. Thus, a sensor network may form a one-hop star network or large networks may form multi-hop networks requiring the use of sophisticated routing algorithms [9] [37]. Though originally developed for military surveillance operations, WSNs today have found their way to civilian applications as well. Typically powered by batteries or from energy harvesting capabilities, a sensor node often tends to be energy constrained and have limited capabilities. The wireless nature, energy constrained operation and widespread uses of such networks open new challenges in the realms of efficient and robust networking with these nodes. A proper framework for analysis of the behavior of sensor nodes in terms of their network operation is an absolute requirement for developing efficient networking protocols. In this chapter we will look into a mathematical analytical tool called game theory, which has been used with considerable success in the field of wireless sensor networks. We will provide a short introduction to game theory and its essential concepts before looking into some of the important advances made in the field using game theory.

Wireless sensor networks present a plethora of problems at the physical, link, and medium access layers. On top of these layers, network and transport layer are there. Besides these layers, there are issues or problems specific to WSNs such as topology control, coverage, localization, time synchronization, etc. Algorithms and protocols for these specific activities are sometimes referred to as helper protocols.

Finally, security in wireless networks is a bigger challenge compared to wired networks due to the inherent shortcomings of the wireless media. No application running on wireless sensor networks can be complete without a proper security analysis. We have chosen some relevant topics or important problems. We show how game theoretic framework has helped us in obtaining good solutions for those problems.

In Section 6.2 below, a brief introduction to game theory and the associated concepts have been given. Section 6.3 presents the problem of channel contention control at the medium access layer and discusses various game theoretic solutions. This is followed by a discussion on the formation of clusters in hierarchical wireless networks in Section 6.4. The solutions offered through game theory are also discussed. Section 6.5 deals with the deployment of sensor nodes using game theoretic approach for countering coverage and connection problems. Jamming, selective forwarding, and other standard security and reliability issues are covered in Section 6.6 along with the game theoretic solution.

6.2 Game Theory: A Short Introduction

Game theory, also known as interactive decision theory, is the study of strategic decision making. According to Roger B Myerson [6], game theory is "the study of mathematical models of conflict and cooperation between intelligent rational decision-makers." Note the use of the term rational decision makers, which is one of the fundamental assumption in game theory, A participant of a game is considered rational in the sense that his decision to cooperate with similarly rational participant depends solely on the gain to be made from their action. The player and participant terms are used interchangeably.

Game theory has been found useful in the fields of economics and political science whose participants are deemed to be motivated only by increasing their gains. More recently, game theory has been found valuable for modelling problems in biology and computer science. This section briefly introduces game theory and the important concepts of: Nash equilibrium [24] [25] and Pareto optimality [11].

6.2.1 Game Theory in Brief

A game consists of a set of players, all of whom are deemed rational. The game is played based on rules which dictate the game. The players compete to maximize their own profit, which in game theory terminology is known as the payoff. The set of actions available to the player from which an action could be chosen to maximize his profit is called the action set. The actions are strategies chosen by the players. The utility is a measure of preferences for some set of actions over the others.

There are many classifications of games. Some relevant ones are given in the following.

6.2.1.1   Cooperative and Non-Cooperative Game

Cooperative games allow players to communicate amongst themselves and form coalitions. Non-cooperative games are played between individuals who do not form coalitions.

6.2.1.2   Zero Sum and Non-Zero Sum Game

A game in which the gain (or loss) incurred by one player is balanced by the loss (or gain) by all other players is known as a zero sum game.

6.2.1.3   Repeated Games

A repeated game is played in multiple stages, where the action of a player in previous steps has an impact on his current and future actions. A game played in a single stage is called a non-repeated game.

6.2.1.4   Complete and Incomplete Information Game

If each player is aware of the possible payoffs for all possible strategies of every other player, they are playing a complete information game. The prisoners' dilemma, detailed below, is an example of a complete information game, where the prisoner (player) is aware of the possible reward or punishment the other prisoner will receive for his action. An incomplete game is one where possible payoff information for other players is not available.

6.2.2   Prisoner's Dilemma

Prisoner's dilemma [2] is a standard example of a non-cooperative game which models the situation where two rational individuals might not cooperate even if cooperation is in their best interest. The game originally framed by Merrill Flood and Melvin Dresher was formalized and named by Albert W. Tucker.

Two prisoners have committed a grave crime and are kept in separate prison cells where they cannot communicate. The police tries to extract a confession from any one or both of them to ease their investigation. The police interrogates them separately and asks them to cooperate with the investigation by confessing to their crime. As a reward for their cooperation, the police agrees to a reduced sentence to the confessing prisoner as his reward. Thus, if both prisoners confess, both get a lighter sentence. If only one of the prisoners cooperates, the confessing prisoner gets the reward of reduced sentence but the other prisoner gets a much harsher sentence. If neither cooperates, police must let both the prisoners walk free due to lack of evidence. Because of its simplicity, prisoner's dilemma is used to introduce many concepts in game theory.

6.2.3   Nash Equilibrium

A Nash equilibrium is a set of strategic choices made by players and their corresponding payoffs in a non-cooperative game. A player will not gain anything by changing his strategy when other players in the game do not change their strategy. There is no guarantee that Nash equilibrium will exist for every game and a Nash equilibrium does not guarantee that it is the most desirable outcome of the game, which is decided by the Pareto optimality. In the prisoner's dilemma game, Nash equilibrium is the strategy when both the prisoners confess to their crime.

6.2.4   Pareto Optimality

Pareto optimality is the allocation of resources such that it is not possible to make one of the player well off without leaving the other competing player worse off. A Pareto improvement is the allocation made after an initial distribution of resources, such that one participant can be made well off without having to leave the other participant worse off. When no further Pareto improvement is possible, allocations are defined as Pareto optimal.

6.2.5   Use in Computer Science

Game theory has found increasing use in Computer science and Logic since logical theories were found to have basis in game semantics. Some of the key areas are like algorithmic game theory [26], algorithmic mechanism design [27] and design and analysis of complex system with economic theory [12].

6.3   Application of Game Theory in Channel Contention Control

The medium access control layer of any network is associated with sensing the common medium (the wireless channel in case of wireless network) to determine whether it is in use by any other node sharing the medium. The data transmission follows only when the channel is sensed as free and the node "acquires" the medium. While IEEE 802.11 [1] is the industry standard in case of wireless networks and IEEE 802.15.4 [36] is targeted for low data rate, low power consumption and low cost networking, several modifications have been proposed in this area. We will briefly look into the three sub-areas of channel assignment, energy efficiency, and contention delay minimization, reviewing a few interesting proposals.

The multi-channel assignment problem has been addressed in the work of Qing yu et al. [44] using a game based channel assignment algorithm in which each player chooses the channel that maximizes its payoff in accordance with the choice of channels by other players, a strategy termed best response. The work by D. Yang et al. [42] proposes a solution to the multi-radio, multi-channel scenario where multiple collision domains are possible by formulating a strategic game modeled on the problem. Another recent but interesting work was [13]. The aim of the work is to reduce power consumption and wireless sensor network interference by modeling a non-cooperative game and finding a MAC algorithm to achieve the objectives.

Energy efficiency is an important domain in wireless sensor networks because of the resource constrained nature of wireless sensor nodes. Another paper by [32] also uses non-cooperative games to optimize performance for protocols using the contention prone MAC in WSNs. The work of Zhao et al. [46] and Mehta et al. [21] deserve special mention.

Another interesting field is contention delay minimization. Contention is the process in which a node tries to acquire the channel on which to send its data. Network efficiency improves when delays from contention is as minimum as possible. Moreover, certain applications of WSNs require nodes to send messages on a priority basis. Contention needs to be as low as possible for them. The work of Francesco Chiti et al. [7] depicts an example of a wireless body area network requiring contention delay minimization. A similar but improved proposal presented by Misra et al. [22] uses a constant model hawk-dove game to prioritize energy data reporting critical scenarios. Other works [8] [31] [32] [33] propose non-cooperative games to optimize protocol performance of shared media using the contention method. The games cited ensure that a unique non-trivial Nash equilibrium is present.

6.3.1 Game-Based Energy Efficient MAC Protocol (G-ConOpt)

In the context of wireless sensor networks, energy efficiency assumes the primary optimization goal that surpasses the traditional performance goals like throughput and delay. Considering the importance of energy efficiency, the work of [46] presents an energy efficient MAC protocol. [46] uses an incompletely cooperative game theory to provide a simplified game theoretic constraint optimization scheme called G-ConOpt, which aims to improve performance at the MAC level while not sacrificing energy efficiency.

Game theoretic analysis: For a node, the game starts with the arrival of a packet and lasts until it moves out of its buffer due to successful transmission or is discarded. The game process is divided into different time slots. During each time slot, each player estimates its present game state according to its earliest history of the game states. Depending on this game state, each player adjusts its strategy which in this case is to tune its contention parameters. Although the player is not aware of the current strategy of its opponents, a prediction can be made from the history.

In the context of this game two players are considered and the node in question and the rest of the nodes are considered as opponents. The present work has taken a different approach to reduce the complexity that would arise from having more than two players. Each player chooses a strategy that maximizes the utility function of the other player. However there are limits to the optimal utility gained by the players. This leads to what is known as a constrained game.

The node being considered, called player 1, can choose from three strategies: transmit, listen or sleep. These correspond to the three contention states for the node. For player 2, the strategies include successful transmission, failed transmission, listening and sleeping. Each node has different payoffs for sleeping, listening, successful transmission or failed transmission. The strategies for the players can be devised from these payoffs, transmission probabilities, probability that the node is sleeping, and probability that a collision will occur.

The simplified game-theoretical constraint optimization scheme: The detailed strategy available from the theoretic formulation above is unfortunately a computationally intractable problem and is inefficient. Thus, a simplified version named G-ConOpt was presented in [46] to optimize performance in limited energy consumption scenarios.

Under this scheme, time is divided into super frames, which are further subdivided into active and sleeping parts. While in the active part, the node contends for the channel in terms of the incompletely cooperative game. While in the sleeping duration, the node turns off its radio to preserve energy. These two durations are determined according to the game state.

The game first estimates the current state, based on which a frame collision probability could be computed. Second, the node adjusts the minimum contention window according to the number of opponents by multiplying the number of opponents with a random number generated. After the end of each game process, the minimum contention window can be determined. If the last transmission was successful, the maximum of nominal minimum contention and half of the final contention window used is selected. If the last transmission was unsuccessful, the maximum contention window is selected.

In IEEE 802.11, the contention process always begins with the nominal minimum contention window; this tends to increase collision in a busy network when the value of contention window is increased. G-ConOpt, recognizes this drawback and improves upon it by using the strategy outlined above. Another adjustment that G-ConOpt makes is to change the duration of active part and the sleeping part. This is accomplished by comparing the number of competing nodes to a predetermined value. If many nodes have frames to send, and the active part is doubled but not exceeding a limit and thereby halving the sleep part not going below a lower bound. If number of competing nodes is less than a predetermined value, it is assumed that few nodes have frames to send and sleep period is doubled up to an upper bound, and subsequently the active period is reduced by half but not below a lower bound.

The performance of the protocol proposed was compared through simulation against S-MAC and CSMA/CA MAC protocols. The results are as follow.

System throughput: G-ConOpt showed a steady rise in throughput as time progressed, levelling out at 25 seconds. CSMA/CA rose at almost the same rate only to level out at 20 seconds instead of rising to 25 seconds. For S-MAC, rise levelled out at 10 seconds. In terms of delay, G-ConOpt fared better, with the CSMA/CA performing similarly until 22 seconds, at which time the delay began to rise more than G-ConOpt. S-MAC delay remained consistently higher than those of G-ConOpt and CSMA/CA. In terms of packet loss rate, the rate is almost zero for G-ConOpt and CSMA/CA, which is lower than that of S-MAC. S-MAC performs better in terms of power consumption that is almost half of that of CSMA/CA. Both these protocols exhibit almost steady power consumption over time. For G-ConOpt, power consumption starts at a low point and increases steadily over time. However, power consumption remains lower than CSMA/CA at all times. In terms of energy efficiency, i.e., ratio of nth rate of successfully transmitted bit rate to power consumption, G-ConOpt fares better than both S-MAC and CSMA/CA. S-MAC starts with better energy efficiency over CSMA/CA; however, over time its energy efficiency drops and becomes almost equal due to increase in traffic load over time.

6.3.2   Game Theory-Based Improved Backoff Algorithm for Energy Efficient MAC Protocol

Another work worth mentioning in the domain of energy efficiency at the MAC level using game theory is [21], which achieved this objective by the use of an improved backoff algorithm developed from an incompletely cooperative game theoretic analysis.

Game formulation: The incomplete game in this case is similar to the one described in G-ConOpt [46]. The payoffs, strategies, and utilities in this case are defined in a similar manner. The approach in [46] was to adjust the contention window based on game theory. The algorithm was found to be computationally non-optimal and had to be simplified greatly to make the algorithm tractable. Other similar approaches were found to be computationally inefficient.

The present work tries a different approach. It introduces an improved back-off mechanism in MAC for energy efficient operation. The work advocates the use of a fixed size contention window but the transmitting probability is non-uniform and should exhibit geometric increase instead of the CSMA/CA binary exponential back-off procedure. In the improved back-off (IB) scheme proposed, the contention window is kept fixed at a small value. The nodes are allowed to choose a transmission slot anywhere between 1 and the value of contention window in a non-uniform, geometrically increasing manner.

It is worth noting that a higher slot number has a better probability of being selected in this scheme. The nodes begin by selecting higher slots in their contention window and starts sensing the channel. If a starting slot has no transmitting nodes, each node adjusts its estimation of competing nodes by multiplicatively increasing its transmission probability in the next slot selection. The process gets repeated and this tends to make competition occur at geometrically decreasing values of an active node number within the contention window considered. Unlike IEEE 802.11, which uses timer suspension, the present method has no such mechanism to meet the objectives of energy savings and latency reduction in case of collision. This may be unfair to the node in question, but the overall aim is to improve performance of the entire network rather than a node. Moreover, in an unsaturated traffic conditions, not all nodes have data to send at all times.

The improved back-off procedure is compared with the normal MAC protocol and an incomplete game protocol in terms of channel efficiency, energy efficiency, and medium access delay in MATLAB®. For channel efficiency, normal MAC exhibits better throughput when the number of nodes is low. However, channel throughput decreases with increase in number of nodes. The improved back-off MAC maintains high channel efficiency due to its approach to collision avoidance. In the incomplete games version, channel efficiency remains almost constant after 30 nodes. In case of medium access delay, delay for normal MAC is more than the other two as more collision occurs. This is followed by improved back-off method. The incomplete game performance was the best. In terms of energy efficiency, the authors claimed that normal MAC wastes more energy due to collision and retransmission attempts. The incomplete game MAC and improved back-off mechanism performed better, giving almost comparative performances.

6.3.3   Game Theory-Based Efficient Wireless Body Area Network

A wireless body area network (WBAN) is a network of sensor nodes utilized to monitor different health parameters of the human body and communicate with a central processing entity for data aggregation and processing. Such networks, on sensing some critical scenario, need to immediately communicate with their central processing entity for necessary action. Undoubtedly, minimizing message delivery latency is of paramount importance. The work presented in [7] considers an improvement of the CSMA/CA protocol with the help of game theory to minimize message delivery latency.

We are aware that game theory finds its use in the analysis of scenarios where participants are deemed rational. An interesting trait of rational entities is that they tend to act selfishly by obstructing the rights of others to maximize their own gain. The idea facilitating one node by depriving others is the backbone of this work. The node which has sensed a critical scenario and needs to send messages fast is allowed to deviate from normal CSMA/CA operation. It is permitted to keep its back-off exponent to the minimum value while others increment that parameter. This is called single cheater node scenario and the node getting preferential treatment is the cheater node. In the single cheater node scenario, delay is reduced for the cheater node but there is no variation in delay of other nodes. Also, collision probability remains unchanged and there is no effect on the access probability of the cheater node and other normal nodes. Thus, in the single cheater model, the goal is accomplished. The proposed protocol advocates only one node to behave deviantly. The effect of allowing more than one cheater node at the same time can be analyzed using the multiple cheater model.

We define each node in the network as a player executing a strategy which consists of choosing the contention window between 1 to maximum window size. The average throughput is considered as the utility function and access probability of each player is considered. Among the players, i.e., nodes in the network, more than one of them may be cheater node and compete for preferential treatment.

A strategy profile is maintained consisting of the strategy chosen by all the players. Two different access probabilities are defined: access probability for the normal nodes and access profile for the cheater nodes. From the analysis, given a generic Nash equilibrium, we can conclude that there exists at least one player with a contention window equal to 1 while others' contention windows are greater than 1. The analysis proves that delivery delay gets unbounded causing overall throughput to drop in a multiple cheater model.

All theoretical claims have been also demonstrated through simulation analysis, using the OMNeT++ mobility framework tool. Both saturated and low loaded conditions have been investigated. In terms of average delay, the cheater nodes have been shown to experience marked reduction in latency irrespective of the number of nodes. Cheater nodes were found to have further low latency with increases in load factor. In terms of dropping probability, there is an overall increase but in low loads there does not seem to be any advantage to the cheater nodes over other nodes. As load reaches saturation, cheater node shows gains over other nodes. Moreover, the theoretical and simulated value of dropping probability has been compared and reported to be in close agreement.

6.4   Application of Game Theory in a Clustered WSN

Clustering is the process in which a group of nodes organizes into groups called clusters and elects a node known as cluster head that has special responsibilities. The non-cluster head nodes send all their data to the cluster head node instead of communicating directly to the base station. The cluster head aggregates all data and sends the summary data to the base station in fixed intervals.

Cluster heads have extra duties and perform more work than other nodes; as a result, they expend more energy. Nodes have selfish interest and may not wish to become cluster heads. A network must enforce the cluster head requirements strictly so that it can continue to function. However, a network must ensure cluster heads are not overburdened with responsibilities and are not selected repeatedly. Fair and equitable distribution of the cluster head responsibility is critical and a network must include energy efficiency measures to ensure equitable load distribution.

This section reviews some of the work in the field of cluster head selection and equitable load distribution in cluster-based WSNs. Koltsidas and Pavlidou [16] proposed a mechanism that uses game theory to manage cluster head selection because nodes acts selfishly and would resist selection as cluster heads unless forced to do so. A refinement to this work [40] utilized local considerations instead of global methodology [16] and the approach proved more effective. Both techniques ensure that cluster head responsibility is evenly distributed across all nodes.

Other works [10] [38] are also worth mentioning. A repeated game model has been developed along with a limited punishment mechanism to discourage nodes from behaving selfishly [38]. The other study [10] presents a game theoretic optimization algorithm that considers the distance from node being selected as a cluster head from other nodes and the remaining energy of the selected node.

Xu et al. [41] suggest a similar solution with a different method. All these proposals consider energy efficiency as a component of cluster head selection. Lin and Wang [35] suggest a game that balances energy of a node by modelling packet transmission. The mechanism includes the imposition of penalties to discourage node selfishness; it also satisfies delivery rate and delay constraints. Jing and Aida [15] provide another solution. These techniques consider the effects of energy efficiency on cluster head selection. Tushar et al. [39] studied power control in a game theoretic framework. The game allows each sensor node to choose its transmission power independently to attain a target signal-to-noise ratio at the receiving cluster.

Hotspots (locations where nodes die off quickly due to overload) are problems in most wireless network; the most catastrophic hotspot is network partitioning. Yang et al. [43] divised a distributed clustering approach using game theory to balance energy consumption throughout a network and avoid hotspots. Their algorithm also adapts cluster size based on the game theory. Salim et al. [30] used the cost and reward mechanism of game theory to achieve cluster head selection.

6.4.1   Game to Detect Selfishness in Clusters (CROSS)

One work [16] concentrated on the clustering problem in the framework of game theory. The analysis is based on the non-cooperative game approach, which models the nature of sensor nodes that behave selfishly to conserve their energy. The game theoretic study and analysis were used to formulate a clustering mechanism called CROSS that can be applied to WSN in practice.

The clustering game consists of sets of players and strategies and a utility function. The players are nodes in the sensor network. The players choose from two strategies: declaring themselves to be cluster heads (CHs) or not. The payoffs are:

1. If a node does not declare and no other declares, payoff is zero.

2. If a node does not declare and some nodes declare, the payoff is a gain value v.

3. If a node declares to be CH: the payoff is $v - c$, where c is the cost incurred to be a CH.

These three conditions define the utility function. The following propositions can be made:

- *Proposition 1:* For a symmetrical clustering game, Nash equilibrium does not exist for the strategy that all nodes declare as CHs.

- *Proposition 2:* For a symmetrical clustering game, Nash equilibrium does not exist for the strategy that all nodes do not declare as CHs.

- *Proposition 3:* Nash equilibrium exists for the strategy where a single node declares itself to be non-CH while all other nodes declare themselves as CHs. Thus, their Nash equilibrium is equal to the number of nodes in the network.

- *Proposition 4:* For a symmetrical clustering game, no symmetric pure strategy Nash equilibrium exists.

- **Theorem 1:** For a symmetrical clustering game, a symmetrical mixed strategy Nash equilibrium exists and the equilibrium probability that the player declares itself as a cluster head depends on the cost incurred to be declared CH and the gain value.

The clustering mechanism is based on the game theoretic study and the natural incentive or cooperation as analyzed previously. The sensor nodes announce their existence in the network. Since the cost incurred to be the cluster head and the gain value are known to every node, the probability to become cluster head could be computed at the first round. This random procedure ensures some nodes declare themselves as CHs and send such beacons. The nodes choose proximity conditions and normal data transfer operations according to clustering protocol. In the second round, the nodes that have not served as CHs are considered for the responsibility, following the zero probability rule. This round-based sequence is a repeated game. When all the nodes have served as CHs, the game resets to its initial condition. Nodes whose energy is depleted are excluded. The clustering mechanism is known as clustered routing of selfish sensors (CROSS).

CROSS performance was analyzed by a comparison with LEACH protocol. Regarding network lifetime, LEACH protocol performs consistently regardless of the cost-to-value (c/v) ratio to become CH.The network lifetime using the CROSS protocol is shorter when the c/v is 0.05. The lifetime is better than that of LEACH at c/v of 0.1 and increases until it reaches 0.5. Lifetime decreases at 0.7 through 0.9 but is still significanly better than that of LEACH. The maximum node lifetime is constant for LEACH systems. CROSS values decrease at an almost constant rate with increase in c/v and are always lower than those of LEACH.

LEACH also demonstrates constant number of clusters. CROSS values decrease steeply at c/v values of 0.05, 0.1, and 0.3, then decrease consistently. CROSS always has fewer clusters than LEACH.

The number of live nodes based on the number of rounds decreases steadily in LEACH systems. The CROSS algorithm at c/v exceeds that of LEACH until it reaches 2250 rounds, after which its c/v is less than that of LEACH. The CROSS algorithm with c/v at 0.9 exhibits stable behavior to 2000 rounds. The c/v drops slightly at 2100 rounds but still remains higher than the LEACH value of 0.1.

However, at 2100, it exhibits sharp decline going below CROSS (c/v=0.1) at 2200 rounds. LEACH also shows decline after a few rounds. CROSS algorithm with computed probability that maximizes the expected payoff is denoted as CROSS2. When network lifetime is compared with number of nodes, CROSS2 has less network lifetime as compared to LEACH for smaller number of nodes. With rise in the number of nodes, network lifetime improves for CROSS and CROSS2. At high communication range, it has been found that network lifetime for CROSS algorithm equals that of LEACH.

6.4.2 Local Game-Based Selfishness Detection in Clustering (LGCA)

The CROSS algorithm uses a game theoretic approach to determine cluster heads in WSNs. This algorithm needs global information on the number of nodes in the network. Global information is difficult to get and it involves sending more packets and consumes more energy. The scheme proposed by [40] improves upon the CROSS mechanism to formulate a localized game theoretical cluster algorithm (LGCA).

The assumption made in CROSS that every sensor node can hear transmission from every other sensor is unrealistic since every node has limits to its transmission distance. Also larger communication range means spending more energy. Another problem is with equilibrium probability. The probability is calculated as an exponentially decreasing function of N (number of nodes) and hence requires large value of N. This results in the sensor nodes having very low probability to become a cluster head, a factor not taken into account by CROSS. The last problem observed by the authors is that the equilibrium average payoff is less than the optimal average payoff.

The LGCA approach consists of an initialization phase followed by many game rounds. Each game round has a set-up phase and a steady state phase.

1. **Initialization phase:** It is assumed that maximum power level of each node is large enough to transfer packets to the base station. Another assumption is that nodes can adjust their power level up to a certain communication distance. In this phase, local neighbor information is collected.

2. **Set-up phase:**

   (a) *Potential CH election:* The probability formula of CROSS is used after replacement of N number of nodes by $Nn(i)$, the number of local nodes. Based on the probability value using the same mechanism as CROSS, the nodes contend to be CH, though the node selected is marked as a potential CH and has to contend with other potential CHs to be selected as the real CH. Similar zero probability rule, as in CROSS algorithm, applies.

   (b) *Real CH contention:* The decision on which a potential CH will be made real CH is determined by the CSMA/CA protocol. The potential CHs contend for channel acquisition based on CSMA/CA and the one that acquires the physical medium is chosen as the real CH and others relinquish their claims as CH.

   (c) *Cluster formation:* When all the CHs have been decided, the CHs broadcast their selection message. The remaining nodes join the nearest CH to form clusters. The nodes then wait for the CHs to allot time slots to them and accordingly utilize the channel.

3. **Steady state phase:** Similar to data collection, aggregation, and their onward transmission to the base station, this phase is like steady state phase of any other cluster algorithm in WSNs.

The performance analysis was performed in MATLAB®. LGCA was compared with the CROSS and LEACH protocols. In the case of network lifetime, LGCA outperforms LEACH and CROSS. With increase in c/v value (defined in CROSS), both CROSS and LGCA experience drops; for CROSS the drop is very steep. Network lifetime has been compared in terms of maximum node lifetime and corresponding number of nodes alive at each round. LEACH achieves maximum node lifetime. However, its energy consumption rate is uneven. The result was confirmed in a comparison of live nodes versus rounds. LGCA exhibited uniform energy expenditures and its maximum node lifetimes exceeded those of CROSS. The average number of cluster heads in CROSS systems decreased sharply as c/v increased. LEACH maintained constant number of clusters. LGCA numbers were relatively constant but displayed slight deviations. So, we can conclude that c/v had little impact on performance of the LGCA protocol.

6.4.3   Game for Cluster Head Selection

Xu et al. [41] proposed a game theoretic approach which aims to restrict the selfish behavior of nodes and promote efficient cluster head selection.

Game theoretic model: A CH should be chosen based on the highest residual energy. However, a selfish node may dishonestly report its residual energy value to avoid being selected as a CH node. Thus, the selfish activity of the node and choice of a CH can be modeled as a game using sensor nodes as the players choosing between either declaring to be CH or refusing to declare. The cost for each strategy is computed by comparing the energy lost by declaring to the energy saved by not declaring. The utility is the difference between payoff and the cost of declaring or not declaring. If no node declares to be CH, the utility for all nodes is zero. However, studies show that at least one node will declare itself as CH.

Cluster heads are important network elements and redundancy is maintained to improve reliability. Apart from the current CH, another node is selected as a candidate CH. The candidate CH replicates data of the current CH and acts as a backup for the sink should the CH fail. To encourage nodes to be candidates, extra payoff is given if a node has a direct communication link with sink.

The game theoretic concepts are used to encourage the nodes to receive the enhanced payoffs and serve as candidates. However, data replication and added communication cost must be considered. As these costs tend to drain energy from nodes, residual energy becomes an important factor. Thus, the candidate nodes must be the least costly in terms of communication and data replication cost, but also should have highest residual energy.

Simulation was carried out in NS-2 to ascertain performance of the proposed approach. The simulation used 100 nodes in an area of 500×500. Throughput was measured. In case of no data replication the throughput dropped to zero after the link between CH and sink dropped. However, with data replication, the throughput resumed its earlier value after the CH sink link goes down and data transmission occured through the candidate-sink link.

6.4.4   Game for Energy Efficient Clustering

A cooperative game theoretic clustering algorithm for WSNs helps to balance energy consumption and improve network lifetime. Jing and Aida [15] proposed a game theoretic model of a clustering algorithm (CGC) for feasibly allocating energy cost to nodes.

The basis of the cooperative game theoretic approach is that a sensor node (SN) trades off individual node costs to the cost of the entire network. A candidate cluster head (CCH) cooperates with other capable SNs to form a cluster head coalition. The parameters to keep in mind are the number of SNs in a cluster, redundancy, and transmission energy. The following conditions apply to the algorithm (described later).

1. Cooperate with another SN which is capable of forming coalition with redundant energy.

2. The system energy consumption is greater than the total cost allocation for CCH node and CCH node with redundant energy.

3. Cooperate with SN in long distance.

4. The system energy consumption is greater than the total cost allocation for the CCH node and the CH node with a distance from CCH.

The algorithm functions as follows

1. At the beginning of round r, each SN elects itself to be a CCH with a probability calculated by considering the residual and initial energy among other parameters.

2. The CCH broadcasts an advertisement message via the CSMA MAC protocol. This allows other SNs to choose the optimum cluster as per the received signal strength.

3. Each non-CCH node sends a join message which includes its identification, residual energy, and distance from CCH.

4. After receiving the join messages of non-CCH nodes in the clusters, the CCH adjusts the final coalition according to the conditions mentioned above.

5. The CCH broadcasts identifications of CHs. The other SNs listen for the CH coalition message.

6. The selected CH broadcasts the decision to other SNs in the network.

7. The non-CHs wait for CH announcements and choose optimum clusters.

8. Each non-CH sends a join message to the CH chosen through received signal strength.

9. After all join messages are received, a TDMA schedule is allocated to each node in the cluster by the CH.

10. The schedule is sent to each constituent node in the cluster and followed by SNs while transmitting the data.

11. The CH collects individual data from each non-CH and sends a summary to the base station.

The performance of the proposed CGC algorithm was compared with LEACH and EEDBC protocol using the NS-2 simulator. The scenario utilizes the simplex energy, random, lattice, semi-lattice, and normal positional distribution. Results of statistical analysis of network lifetime and energy efficiency used a confidence interval of 0.975. The simulation results show that CGC and EEDBC prolonged network lifetime by 25%. The present algorithm also outperformed LEACH and EEDBC by 24.5% and 21.6% respectively. CGC also reduced data transmission latency; the simulation also demonstrated that CGC was the most efficient scheme for data transmission per unit of energy. CGC improved network lifetime and data transmission capacity by up to 5.8% and 35.9%, respectively. The simulation results show that the CGC routing in WSNs extends the network life, reduces transmission time, and regulates clustering to achieve energy efficiency.

6.5 Application of Game Theory in Connection and Coverage Problem

Considering the resource constrained nature of the sensor nodes in wireless sensor networks, maximizing the coverage of nodes and maintaining connectivity is not a trivial problem. The coverage area of nodes in such cases could be analyzed and effective algorithms could be developed using a game theoretic framework. The survey below describes some of the work where game theory has been applied to improve network coverage and ensure connectivity in wireless sensor networks.

The work of Zhang et al. [45] will be described later. Senturk et al. [34] describe restoration of connectivity by positioning relay nodes to mitigate network partitioning caused by node failures. For the strategy to work, the number of partitions, damaged area, and locations of remaining nodes must be known. Coverage control is the concept of prolonging network lifetime while meeting user's objectives. The K-cover algorithm is an accepted solution. It has the usual disadvantages of centralized systems [14] but uses a game theoretic model to achieve the dual objective of network lifetime maximization and widest coverage.

6.5.1 Network Coverage Game

The algorithm proposed by Zhang et al. [45] aims to improve network coverage in WSNs using game theoretic concepts.

Game formulation: A repeated game model has been developed. The set of nodes in the network represents the set of players in the game which may adopt the strategy of increasing, decreasing or following their energy cover area. A player can revise its strategy until the final stage. The energy coverage area achieved is the final solution. The optimal solution to the game is linked to finding its payoff. Certain definitions of payoff are important.

Coverage: Coverage is the ratio of the actual cover area to the required cover area.

Coverage level: Coverage level is the number of nodes in the whole network that can cover a concerned node.

Now, we can define the payoff of every node by considering the following:

1. When the value of complete coverage is 1, complete coverage is the most important factor for all nodes.

2. When the value of a node's energy cover area is determined by the product of the square of inductive radius and a negative of the small value deciding the value of energy consumption, we can infer that small energy cover areas are preferred.

3. The value of the node's inductive area edge location is dependent on the product of level efficiency of coverage level and a small parameter whose value decides the importance of coverage level to be considered during a node's decision. Coverage level of all nodes in the circle must be considered. If the efficiency value is increased in a position, it means the covered position is not important and the overall efficiency decreases and vice versa.

Thus, the whole network's payoff is the sum of payoff of all the nodes in the network. That is, payoff is the sum of the three formulations described above. The algorithm works as follows:

1. Initialize the game number and number of nodes to zero and begin iteration.

2. For each game follow the steps below until all game stages are played.

3. For all nodes:

   (a) Play the game

   (b) Choose strategy

   (c) Choose action

4. At the end of each game, change the network.

5. At the end of all the games, confirm the network set-up.

Further improvement can be brought about by the use of anticipant nodes. A node in a position is the anticipant node. In the previous algorithm, every node considers the use of minimal energy and covers neighbor position. In this improved scheme, we decide whether nodes should consider this position or the anticipant nodes in a position. When a position is out of consideration, efficiency of coverage level is decreased and vice versa.

The proposed mechanism and its improvement are simulated in VC++ to measure their effectiveness. There are 100 nodes distributed randomly in an area of 100×100 sq. meters. When only GCC is evaluated with repeated game, inductive areas are adjusted and network coverage comes to 100% with many sleeping nodes. An evaluation of improved GCC shows more sleeping nodes and complete coverage is obtained. On comparative performance evaluation, we see an optimization of 40% in the use of improved GCC. Thus, with a change of payoff, huge optimization is possible.

6.6 Application of Game Theory in Security and Reliability of WSNs

A wireless node in a wireless sensor network is always resource constrained. Thus, a selfish behavior is the natural rational choice for such nodes. However, selfish behavior and resulting non-cooperation result in adverse impact on network performance. Game theory finds widespread application in resolving such issues arising from selfish behavior of nodes and helps in proper functioning of the network.

Outright malicious behavior like jamming, selective forwarding, and active denial of service can also be tackled by the application of game theoretic analysis as we will see in the following text. Amongst the works focusing on malicious behavior of nodes, [3] [19] [28] are described later. Refs. [18] and [20] deserve mention in this category. The work in [20] considers a situation where a malicious node may jam the network by sending an abnormal data packet to another node to block the channel from doing anything useful.

The remedial approach proposed in the work is what is known as the pairwise simultaneous game, where a game is played between two nodes in the network. The node pair may be normal-normal, normal-malicious, or maliciousmalicious. The nodes do not know each other's identity, i.e, whether malicious or normal. Ref. [18] considers a situation where a malicious node broadcasts a jamming signal to interfere the working of a wireless smart grid network.

In this case, the problem of jamming and the solution of anti-jamming have been modeled as a zero sum stochastic game. The Nash equilibrium strategy is analyzed, as claimed in the paper, to provide better performance. Amongst the works about selfishness in WSNs, apart from the work by [5] detailed later, the works of [17] [23] [29] and [47] deserve mention. Ref. [29] considers the wireless random access of non-cooperative selfish transmitting nodes. The authors formulated a non-cooperative random access game in which transmission strategies in non-cooperative Nash equilibrium are derived depending on the parameters of rewards attained through throughput, cost incurred in delay and energy. This analysis was used to formulate optimal strategies to block random access of selfish nodes. Also, a strategy concerning optimal defense against denial of service attack has been devised. Mukherjee et al. [23] address the problem of selfish behavior of nodes in terms of non-cooperative game theory. The nodes choose to forward a packet or not, a characteristic which is analyzed through the game. The authors went on to prove the existence of Nash equilibrium which gives a solution for the game. Zheng's work [47] uses a game theoretic approach to find a balance between resource constraints and reliability through cooperation enforcement. Another method [17] uses evolutionary game theory and a genetic algorithm over AODV known as GA-AODV to model the behavior of selfish nodes and propose solutions.

6.6.1   DoS Game

Denial of service (DoS) attack is one of the most notorious type of attacks on any network system. WSNs are no exception. Different protocols have been proposed for network systems to cope with this menace. A DoS attack lends itself to be easily described as a game between the attacker and the intrusion detection system (IDS). This concept forms the basis of a work embodied in [3], which proposes to prevent a passive DoS attack in a WSN by the use of repeated games.

The Agah and Das [3] proposal is modeled as a non-cooperative N player game between normal and malicious nodes. The IDS residing at the base station or sink monitors node collaborations and rate all nodes based on their contributions to total network activity. Positively rated nodes are rewarded.

The game involves players (nodes in a network), a set of actions (forwarding or dropping packets), and past histories of nodes. The IDS tracks game history and alerts nodes of malicious nodes in their neighborhood. The von Neumann-Morgenstern utility function is considered for each node during each game. Payoff is calculated as the difference between reputation of a node and the cost of transmitting a packet.

The nodes respond by cooperating (forwarding packets) or not cooperating. The IDS responds by catching or missing malicious behavior. Four possibilities result from combining the responses of the nodes and the IDS: (1) least damage; (2) false positive; (3) false negative, and (4) best choice for IDS. The four possibilities have different utility values. The IDS initially considers all nodes to act normally. If malicious activity is detected, the IDS assumes the affected nodes will remain malicious for the rest of the game. The IDS builds node reputations over time by monitoring their activities. The reputations are used to assess the trustworthiness of the nodes and predict their future behaviors. A node collaborates only with trustworthy nodes whose reputations are used and their values in the network are based on their reputations. The protocol steps are as follows.

1. A node sends a `Route_request` message.

2. All nodes receiving the message compute their utility.

3. If no previous `Route_request` is received, the nodes place themselves in the source route and forward the `Route_request` to their neighbors.

4. If a destination is reached or a route to destination is available, the `Route_request` message is not propagated further and a reply is sent.

5. After receiving the replies, the source selects the route with highest utility because it has the highest number of trustworthy nodes.

6. Once a route request reaches a destination, the path that the request has taken is reversed and sent back to the sender.

7. When a destination notifies the base station about the receipt of the packet, the IDS is informed and subsequently increases the reputation of every node on the path. The new reputation value is broadcast to the nodes.

8. As each node is aware of its neighbors, it will update the reputation table.

The simulation was carried out with the following assumptions.

1. Sensor nodes are scattered in the field.

2. Each node starts with the same power.

3. Two sensor nodes are able to communicate if they are within the transmission range of each other.

4. Sensors perform their task and periodically report to base stations.

5. IDS is present in the base stations and constantly monitors all nodes for maliciousness.

6. Some malicious nodes are present and may launch DoS attack.

The simulation shows that in absence of any attacker the nodes are able to transmit successfully 60% of the time. With 10% malicious nodes present throughput drops to 52% and with 20% attacking nodes, throughput drops to 35%. Without attack, a packet is received over a mean average hop length of 7. Long routes are preferred for nodes without attacks. However, with attacks the tendency is to send through shorter routes. It has been found that the average throughput of a malicious node decreases with time.

Overall, the simulation concludes that lower the number of malicious nodes in the network, the better its detection ability. With more malicious nodes present, the IDS gets cautious about its own utility which is decreased with false positives and false negatives. It misses some of the malicious nodes to avoid losing utility because of false positives and false negatives.

6.6.2   Jamming Game

The use of IDS is an effective strategy for preventing attacks in WSNs. While mitigating attacks in WSNs improves system efficiency and reduces system cost, IDS is costly as it uses system resources. Thus, it is desirable that IDS is used only when need arises. Ma et al. [19] propose a solution in which the IDS need not be kept ON always. Using a non-cooperative game theoretic framework, it helps cluster head nodes decide the probability of starting the IDS service.

Game model: A non-cooperative game is played with two strategies for the cluster head to start the IDS or not to start the IDS. The attacker has three strategies. The attacker can attack the cluster head in question which invokes the reaction from the cluster head of starting the IDS or not starting the IDS. The attacker can alternately do nothing but wait or it can attack another cluster head. There is no strategy of Nash equilibrium in the game. On analyzing the game theoretic model, the model was found suitable in saving system resources used by IDS to monitor attacks.

The performance of the proposed scheme was analyzed using the GloMoSim testing platform. A jamming attack was used to simulate the attack condition. The simulation result shows that the scheme proposed was able to detect 70% to 85% of attacks after attaining stability. The number of nodes failing to always monitor condition was attained at 237 rounds, while for the game model proposed the all-node-death condition was attained at 541 rounds.

6.6.3   Malicious Node Detection Using Zero Sum Game

Because of the autonomous nature and limited resources of WSNs, security is a major issue. Malicious node detection is one of the primary security concerns in such networks. A framework for malicious node detection using zero sum game approach and selective node acknowledgement in the forwarding path was proposed [28].

In a typical WSN topology, malicious nodes responsible for selective forwarding attack are scattered amongst the normal ones. In a sensor network the aim is to transfer data to the base station. The hacker tries to take control of the routing layer of nodes, thereby disrupting the data flow. In case of cluster formation, the cluster heads are targeted.

Game theoretic formulaion: The consideration is that the IDS responsible for defending against the attack is at the cluster head and monitors the data transfer. The attacker may attack any node including the cluster head. Two terms were defined: Ea, energy of the attacker, and Ec, energy of the IDS node.

The following conditions determine whether an attack will succeed:

1. *$Ea>Ec$:* The attack will succeed.

2. *$Ea = Ec$:* The attack may or may not succeed.

3. $Ea = 0$: There is no attack.

The payoff in the IDS is its utility function U which is the difference between energy of the IDS node and energy of the attacking node.

Based on game theory-based analysis the following important theorems and definitions are discussed.

Definition 1: If the zero-sum game is Pareto optimal, it is called a conflict game.

Definition 2: A game is zero-sum if for all outcomes the sum of payoffs of all nodes is zero.

Definition 3: In a zero-sum game of cost of an attack on a node is equal to the cost of defending it. The payoff of a successful attack by an intruder equals the loss (in energy) by the IDS of a cluster.

Theorem 1: A zero sum game has no pure Nash equilibrium.

Definition 4: The total cost to successfully attack a node (or a cluster of nodes) depends on the number of attacks.

Definition 5: The IDS defends many intruder attacks. For a zero sum game, the cost of the intruder's success and failure attacks equals the cost of success and failures of defending the nodes.

Theorem 2: Total cost to defend (energy required) a cluster is constant. In other words, the zero-sum game reaches Nash equilibrium when an intruder attacks the cluster and the IDS defends it.

Important corollary: The zero-sum game reaches Nash equilibrium at a malicious node.

The detection mechanism: Ideal radio condition is assumed. Thus, packet drops are due to malicious activity by nodes. Assume that m nodes are malicious out of n nodes in the forwarding path. Let q be the number of non-malicious nodes, k, the number of selected acknowledgement points, and a be the percent of nodes randomly selected as checkpoints. The probability of detecting a malicious node is the probability of packets dropped by selected acknowledgement points. The probability of packets being dropped between specified checkpoints is the difference between the probabilities of acknowledgements at the source by the first checkpoint and that of the second checkpoint.

Simulation results performance measurements: The simulation for performance measurement was carried out in MATLAB®. One hundred and twenty nodes were introduced between source and sink and malicious nodes were chosen randomly in the path. For 20 iterations conducted, it was observed that malicious node is one or two nodes away from the selected acknowledgement points. The detection of malicious node was always found between acknowledgement points irrespective of drop rates.

6.6.4   Selective Forward Prevention Game

Asadi et al. [5] proposed a protocol for cooperation enforcement through a reward-punishment scheme where a node in the sensor network rationally decides whether to forward packets coming from other nodes to save its own power. The nodes are assisted in their rational decision making with the game theoretic framework, wherein suitable gains could be made on forwarding packets and a history of selfish behavior of nodes is kept to set them aside from the network.

Game formulation: Available battery power of each node in the network is considered to be the benefit which each node tries to maximize. Thus, nodes have a tendency to refrain from participating in forwarding packets to conserve their energy. Appropriate incentives must be given to these nodes to compensate for their loss of energy by participating in packet forwarding process.

The incentive considered in this case is a reputation value, which represents how trustworthy a node is. Low reputation nodes are deemed malicious and removed from the network.

The game has three components: the players or the nodes in the network, the set of actions or strategies that a player can choose from and a utility function. An equilibrium condition is a state in which two players agree on an outcome in some set X, which is feasible division of reputations, or the outcome will be some fixed event D, which neither party receives a positive reputation.

The two main goals are minimizing battery usage and maximizing cooperation among the nodes. Thus, partial cooperation does not lead to equilibrium. So, the aim is to devise a strategy to enforce full cooperation among the nodes, while also being aware of the power conditions of the nodes to avoid energy loss by nodes attempting to gain reputation through the game.

The nodes have von Neumann-Morgenstern utility functions defined over the outcomes of stages in a repeated game. The nodes have two action choices: forward the packet or not forward the packet. Utility function is defined as the difference between weighted reputation value and the weighted cost incurred by node in forwarding the packet. The weight assigned to the reputation value is the number of units of time since the node last forwarded a packet. The weight assigned to the cost value depicts the importance of conserving energy. Power level decreasing below a certain threshold means node weight value is increased. The reputation is assigned based on the summation of ratings awarded by base stations to nodes. This is called subjective reputation.

Malicious node detection: The base station monitors the node's behavior and periodically checks for maliciousness based on throughput. The current reputation value of each node in the network is considered and an average and standard deviation are calculated. If the reputation is lower than the difference between the average and standard deviation calculated, the node is deemed malicious. The base station informs the node to shut down and close its radio communication.

Network configurations: The following three configurations were proposed.

- *Case 1:* The sensor nodes broadcast to the nodes in a certain range. The nodes receiving the broadcast rebroadcast them to further nodes in their range. However, this creates a lot of packets in the network.

- *Case 2:* Each node has identifications (IDs) of neighbors recorded in the neighborhood table selected by a handshaking protocol at boot up time. On receiving a packet from a node, the ID is checked with the neighborhood table. If ID is not found, no further action is taken. If a match takes place, the node's number of packets received is incremented and appropriate action is taken.

- *Case 3:* The third configuration is based on clustering. A cluster head node is chosen based on the battery power available among nodes.

Three possible configurations are simulated. The simulation results show that the reputations in game theory-based method are lower than in non-game theory-based ones. The network size does not matter. For configuration case 1, game theoretic implementation results in higher voltage loss for larger networks than smaller networks. This is because the decision to forward packets is energy consuming and for larger networks, the decision process needs more energy than required for forwarding packets. For configuration case 2, game theoretic implementation results in consistent lower voltage loss. The neighborhood system generates less traffic than the broadcast system of case 1. For detecting malicious nodes, a lower number of malicious nodes aids in better detection. The detection procedure is based on average and standard deviation; thus, if many nodes have lower reputations, detection becomes difficult. Game theoretic implementation raises fewer false positives. A small number of false positives are generated in case 1. The average reputation for game theoretic implementation is lower than in non-game theoretic ones except for clustering scenarios. In this case, reputation is higher for game theory cases with large percentages of malicious nodes. Voltage loss is lower in game theory cases than in non-game theory ones, even when malicious nodes are present. Average network utility decreases for case 1; implementations using game theory have lower utility than game theory based implementations due to high traffic in the network. For case 2, utility is higher for game theory implementing networks than non-game theory implementing ones, despite a large network size. For case 3, average utility is higher for non-game theory cases than game theory cases.

6.7 Concluding Remarks

Though originally developed as a mathematical analysis tool in the realms of economics and political science, game theory has made its mark in the field of computer science as well. In the sub-domain of computer networks, especially wireless networks, game theory is becoming an indispensable tool for analyzing such networks and creating improved protocols and algorithms. In this survey, we have provided a glimpse of the copious efforts in the field of wireless sensor networks using game theory. From the MAC sub-layer to network layer routing, game theory has become a model of choice because many of the typical situations lend themselves to be modelled as such.

However, the dynamicity and uncertainty in the field of wireless sensor networks has increased the complexity of the application of game theory. Research to tackle such issues is promising. In spite of its complexity, game theory continues to be a tool of choice for researchers and will remain so for the years to come.

Bibliography

[1] IEEE Std 802.11-2007, part 11: Wireless LAN Medium Access Control (MAC) and Physical Layer (PHY) specifications. *IEEE Std. 802.11, 2007*, 2007.

[2] A. Dixit and B. Nalebuff, Prisoner's Dilemma. In David R. Henderson (ed.). *Concise Encyclopedia of Economics*. 2008.

[3] A. Agah and S. K. Das. Preventing DoS Attacks in Wireless Sensor Networks: A Repeated Game Theory Approach. *International Journal of Network Security*, 5(2):145 – 153, September 2007.

[4] F. Akyildiz and I. H. Kasimoglu. Wireless Sensor and Actor Networks: Research Challenges. *Ad Hoc Networks*, 2(4):351367, October 2004.

[5] M. Asadi, C. Zimmerman, and A. Agah. A Game-theoretic Approach to Security and Power Conservation in Wireless Sensor Networks. *International Journal of Network Security*, 15(1), January 2013.

[6] R. J. Aumann. *Game Theory, The New Palgrave: A Dictionary of Economics, Second Edition*. 1987.

[7] F. Chiti, R. Fantacci, and S. Lappoli. Contention Delay Minimization in Wireless Body Sensor Networks: A Game Theoretic Perspective. In *Proceedings of the 2010 IEEE Global Telecommunications Conference (GLOBECOM 2010)*, pages 1 – 6, December 2010.

[8] S. Chowdhury, S. Dutta, K. Mitra, D. K. Sanyal, M. Chattopadhyay, and S. Chattopadhyay. Game-Theoretic Modeling and Optimization of Contention-Prone Medium Access Phase in IEEE 802.16/WiMAX Networks. In *Proceedings of the 2008 Third International Conference on*

Broadband Communications, Information Technology & Biomedical Applications, pages 335 – 342, November 2008.

[9] W. Dargie and C. Poellabauer. *Fundamentals of wireless sensor networks: theory and practice*. John Wiley & Sons, Inc. New York, NY, USA, July 2010.

[10] S. Dasgupta and P. Dutta. A Novel Game Theoretic Approach for Cluster Head Selection in WSN. *International Journal of Innovative Technology and Exploring Engineering (IJITEE)*, 2(3), February 2013.

[11] D. Fudenberg and J. Tirole. *Nash equilibrium: multiple Nash equilibria, focal points, and Pareto optimality in Fudenberg, Drew; Tirole, Jean, Game theory*. Cambridge, Massachusetts: MIT Press, 1991.

[12] J. Y. Halpern. *Computer Science and Game Theory, The New Palgrave Dictionary of Economics, 2nd Edition*. 2008.

[13] X.-C. Hao, Q.-Qian Gong, S. Hou, and B. Liu. Joint Channel Allocation and Power Control Optimal Algorithm Based on Non-cooperative Game in Wireless Sensor Networks. *Wireless Personal Communications*, 78(2):1047 – 1061, 2014.

[14] X. He and X. Gui. The Localized Area Coverage Algorithm Based on Game-Theory for WSN. *Journal of Networks*, 4(10), 2009.

[15] H. Jing and H. Aida. A cooperative game theoretic approach to clustering algorithms for wireless sensor networks. In *Proceedings of the IEEE Pacific Rim Conference on Communications, Computers and Signal Processing, 2009. PacRim 2009*, pages 140 – 145, August 2009.

[16] G. Koltsidas and F.-N. Pavlidou. A game theoretical approach to clustering of ad-hoc and sensor networks. *Telecommunication Systems*, 47(1):81 – 93, June 2011.

[17] K. Komathy and P. Narayanasamy. Secure Data Forwarding against Denial of Service Attack Using Trust Based Evolutionary Game. In *Proceedings of the IEEE Vehicular Technology Conference, 2008. VTC Spring 2008*, pages 31 – 35, May 2008.

[18] H. Li, L. Lai, and R. C. Qiu. A denial-of-service jamming game for remote state monitoring in smart grid. In *Proceedings of the 2011 45th Annual Conference on Information Sciences and Systems (CISS)*, pages 1 – 6, March 2011.

[19] Y. Ma, H. Cao, and J. Ma. The intrusion detection method based on game theory in wireless sensor network. In *Proceedings of the 2008 First IEEE International Conference on Ubi-Media Computing*, pages 326 – 331, August 2008.

[20] Y. Mao, P. Zhu, and G. Wei. A Game Theoretic Model for Wireless Sensor Networks with Hidden-Action Attacks. *International Journal of Distributed Sensor Networks*, 2013 (2013), July 2013.

[21] S. Mehta and K. S. Kwak. An energy-efficient MAC protocol in wireless sensor networks: a game theoretic approach. *EURASIP Journal on Wireless Communications and Networking*, 2010(17), April 2010.

[22] S. Misra and S. Sarkar. Priority-Based Time-Slot Allocation in Wireless Body Area Networks During Medical Emergency Situations: An Evolutionary Game-Theoretic Perspective. *IEEE Journal of Biomedical and Health Informatics*, 19(2), March 2014.

[23] S. Mukherjee, S. Dey, R. Mukherjee, M. Chattopadhyay, S. Chattopadhyay, and D. K. Sanyal. Addressing Forwarder's Dilemma: A Game-Theoretic Approach to Induce Cooperation in a Multi-hop Wireless Network. *Advances in Communication, Network, and Computing*, 108:93 – 98, 2012.

[24] J. Nash. Non-Cooperative Games. *The Annals of Mathematics*, 54(2):286 – 295, 1951.

[25] J. F. Nash. Equilibrium points in n-person games. In *Proceedings of the National Academy of Sciences*, volume 36, pages 48 – 49, 1950.

[26] N. Nisan. *Algorithmic Game Theory*. Cambridge University Press. Description, 2007.

[27] N. Nisan and A. Ronen. Algorithmic mechanism design. In *Proceedings of the Thirty-First Annual ACM Symposium on Theory of Computing*, pages 129 – 140, 1999.

[28] Y. B. Reddy and S. Srivathsan. Game theory model for selective forward attacks in wireless sensor networks. In *Proceedings of the 17th Mediterranean Conference on Control and Automation, 2009. MED '09*, pages 458 – 463, June 2009.

[29] Y. E. Sagduyu and A. Ephremides. A game-theoretic analysis of denial of service attacks in wireless random access. In *Proceedings of the 5th International Symposium on Modeling and Optimization in Mobile, Ad Hoc and Wireless Networks and Workshops, 2007. WiOpt 2007*, pages 1 – 10, April 2007.

[30] S. Salim, X. Li, and S. Moh. Cost- and reward-based clustering for wireless sensor networks: A performance tradeoff. In *Proceedings of the 2013 22nd Wireless and Optical Communication Conference (WOCC)*, pages 421 – 425, May 2013.

[31] D. K. Sanyal, M. Chattopadhyay, and S. Chattopadhyay. Improved performance with novel utility functions in a game-theoretic model of medium access control in wireless networks. In *Proceedings of the TEN-CON 2008 - 2008 IEEE Region 10 Conference*, pages 1 – 6, November 2008.

[32] D. K. Sanyal, M. Chattopadhyay, and S. Chattopadhyay. Performance Improvement of Wireless MAC Using Non-Cooperative Games. *Advances in Electrical Engineering and Computational Science*, 39:207 – 218, 2009.

[33] D. K. Sanyal, M. Chattopadhyay, and S. Chattopadhyay. Recovering a game model from an optimal channel access scheme for WLANs. *Telecommunication Systems*, 52(2):475 – 483, February 2013.

[34] I. F. Senturk, K. Akkaya, and S. Yilmaz. Relay placement for restoring connectivity in partitioned wireless sensor networks under limited information. *Ad Hoc Networks*, 13(Part B):487 – 503, February 2014.

[35] X. -H. Lin and H. Wang. On using game theory to balance energy consumption in heterogeneous wireless sensor networks, *In Proc. of IEEE Local Computer Networks*, 568–576, October 2012.

[36] A. F. Molisch, K. Balakrishnan, D. Cassioli, C. Chong, S. Emam, A. Fort, J. Karedal, J. Kunisch, H. Schantz, U. Schuster, and K. Siwiak. IEEE 802.15.4a channel model - final report.

[37] K. Sohraby, D. Minoli, and T. Znati. *Wireless sensor networks: technology, protocols, and applications*. John Wiley & Sons, Inc. New York, NY, USA, March 2007.

[38] L.-hui Sun, H. Sun, B.-qing Yang, and G.-jiu Xu. A repeated game theoretical approach for clustering in mobile ad hoc networks. In *Proceedings of the 2011 IEEE International Conference on Signal Processing, Communications and Computing (ICSPCC)*, pages 1 – 6, September 2011.

[39] W. Tushar, D. Smith, T. A. Lamahewa, and J. Zhang. Non-cooperative power control game in a multi-source wireless sensor network. In *Proceedings of the 2012 Australian Communications Theory Workshop (AusCTW)*, pages 43 – 48, January 2012.

[40] D. Xie, Q. Sun, Q. Zhou, Y. Qiu, and X. Yuan. An Efficient Clustering Protocol for Wireless Sensor Networks Based on Localized Game Theoretical Approach. *International Journal of Distributed Sensor Networks*, 2013 (2013), January 2013.

[41] Z. Xu, Y. Yin, X. Chen, and J. Wang. A Game-theory Based Clustering Approach for Wireless Sensor Networks. In *Proceedings of the 2nd International Conference on Next Generation Computer and Information Technology*, volume 27, pages 58 – 66, 2013.

[42] D. Yang, X. Fang, and G. Xue. Channel allocation in non-cooperative multi-radio multi-channel wireless networks. In *Proceedings of the 2012 IEEE INFOCOM*, pages 882 – 890, 2012.

[43] Y. Yang, C. Lai, L. Wang, and X. Wang. An energy-aware clustering algorithm via game theory for wireless sensor networks. In *Proceedings of the 2012 12th International Conference on Control, Automation and Systems (ICCAS)*, pages 261 – 266, October 2012.

[44] Q. Yu, J. Chen, Y. Fan, X. Shen, and Y. Sun. Multi-channel assignment in wireless sensor networks: A game theoretic approach. In *Proceedings of the 2010 IEEE INFOCOM*, pages 1 – 9, March 2010.

[45] L. Zhang, Y. Lu, L. Chen, and D. Dong. Game Theoretical Algorithm for Coverage Optimization in Wireless Sensor Networks. In *Proceedings of the World Congress on Engineering*, volume 1, 2008.

[46] L. Zhao, L. Guo, G. Zhang, H. Zhang, and K. Yang. An energy-efficient MAC protocol for WSNs: game-theoretic constraint optimization. In *Proceedings of the ICCS 2008 11th IEEE Singapore International Conference on Communication Systems*, pages 114 – 118, November 2008.

[47] M. Zheng. Game Theory Used for Reliable Routing Modeling in Wireless Sensor Networks. In *Proceedings of the 2010 International Conference on Parallel and Distributed Computing, Applications and Technologies (PDCAT)*, pages 280 – 284, December 2010.

Chapter 7

Optimal Cluster Head Positioning in Heterogeneous Sensor Networks

Biswanath Dey

National Institute of Technology

Sandip Chakraborty

Indian Institute of Technology

Sukumar Nandi

Indian Institute of Technology

7.1 Introduction

A wireless sensor network is a spatially distributed autonomous system of sensor nodes used for health monitoring, environment surveillance, military intelligence and other purposes. A wireless sensor network consists of a base station and a set of distributed sensor nodes. Sensor nodes collect data from

153

their environment and send it to the base station. The base station acts as a sink. Some of the characteristic features of wireless sensor networks such as limitation of resources (processing power, energy) make them different from other distributed networks. Consideration of network lifetime is essential for any protocol.

Some works such as [3, 6, 18, 30] specifically consider homogeneous sensor networks. These networks consist of sensor nodes with similar energy, hardware and transmission capacities. Works such as [3, 9, 17] consider a heterogeneous sensor network (sensor nodes vary in terms of battery powers). Such networks generally consist of a mixture of high energy sensor nodes (advanced nodes) and low energy nodes (plain nodes).

To improve network lifetime, nodes organize themselves into a hierarchical arrangement to form localized groups called clusters. Each cluster consists of a cluster head and several member nodes. The member nodes collect data and send it to their cluster heads. The cluster head aggregates and transmits the data to the base station. The energy consumption of cluster heads is higher than that for member nodes. A heterogeneous energy distribution is proficient for clustered networks. However, a higher percentage of advanced nodes must be elected as cluster heads. The performance of a clustering protocol improves with the election of a higher fraction of advanced sensor nodes as cluster heads.

Routing in a cluster based heterogeneous sensor network depends on several factors, such as remaining energy and energy consumption rates of the intermediate sensors in the routing path, routing path length, cluster efficiency, traffic density and network stability. Based on these factors, this chapter designs a multi-objective vector optimization to find the best forwarding path in a cluster based heterogeneous sensor network. As the solution for a multi-objective optimization is known to be NP-hard, this chapter uses the technique of evolutionary optimization [1, 11, 31] where every sensor node uses an evolutionary approach to maximize its local solution view that in turn finds the global best end-to-end routing paths. The performance of the proposed scheme is also analyzed using simulation results. In summary, the major objectives of this chapter are twofold.

1. This chapter describes a *cluster head positioning problem* (CHPP) in heterogeneous wireless sensor networks, for energy optimized and bounded hop data forwarding.

2. A multi-objective optimization is formulated for cluster based wireless sensor networks with the vector objectives of minimizing forwarding delay while maximizing concurrent packet transmissions. The constraints are modeled based on interference free scheduling and the rate of energy dissipation.

3. An evolutionary approach based on decomposition is designed to solve the multi-objective optimization at every individual sensor node to decide its next hop forwarder towards the cluster head. The performance of this solution approach is evaluated using simulation results.

The rest of the chapter is organized as follows. Section 7.2 discusses related works on cluster head positioning in sensor networks, and gives an introductory idea on multi-objective optimization methods through evolutionary computing and its applications. Section 7.3 formulates the cluster head positioning problem. The evolutionary optimization solution of the CHPP problem is discussed in Section 7.4. Section 7.5 analyzes the performance of the proposed method through simulation results. Finally, Section 7.6 concludes the chapter.

7.2 Related Works

Dynamic cluster head selection algorithms are well studied in the sensor networks literature. Routing and clustering have been intensely studied in contemporary research in wireless sensor networks [3, 6, 8, 9, 16, 18, 19, 27, 34]. In order to cope with the unique characteristics of sensor networks described in the previous section, newer range of protocols had to be developed. Low energy adaptive clustering hierarchy (LEACH) [16] is a hierarchical clustering protocol specific to the sensor network and the first of its kind, which uses clustering for prolonging network lifetime. LEACH has four phases of operation: advertisement, cluster setup, schedule creation (TDMA), and data transmission. In the advertisement phase each node decides whether it can become a cluster head, based on a predetermined percentage P, of a cluster head desired in the network with respect to total number of nodes in the network. With given cluster head probability P, during start of a network round r, a node that has not become a cluster head in past $(r - 1)$ rounds, tries to become the cluster head by generating a random number between 0 and 1, which is compared with a threshold $T(n)$, calculated as shown below.

$$T(n) = \frac{P}{1 - P * (r \times \mod (\frac{1}{P}))}$$

If the random number selected by a node is less than $T(n)$, the node becomes the cluster head. Each node that becomes a cluster head advertises itself using the medium access control based CSMA protocol and transmitting at the same energy. In cluster setup phase each non-cluster-head node decides which cluster to join based on the received signal strength from the cluster head advertisement. This process is repeated periodically with the aim so that every node in the network becomes a cluster head and all nodes can equally share the responsibility of message transmission, ensuring longer life for all the nodes. In schedule creation phase the cluster head builds a TDMA schedule for the member nodes and transmits the schedule to each member node and in data transmission phase each member node transmits its data to the cluster head. At the end of every TDMA cycle the cluster head aggregates all packets and sends another set of messages to the BS as CDMA packets.

Although LEACH is one of the most elegant protocols studied by many researchers [5,26] the randomized nature of its cluster head selection has some limitations. There is ample scope for improvement as far as the network lifetime and energy efficiency are concerned as described in [5]. Efficient scheduling of cluster head selection biased by node location and expected network lifetime can increase the network performance which is the core focus of our research.

A major improvement over LEACH was achieved by PEGASIS [25] for network lifetime, which is claimed to be the near optimal solution by its inventors. In PEGASIS, group heads are chosen randomly and each node communicates only with close neighbors to form a chain leading to its cluster head. The randomization of the cluster heads guarantees that the nodes will die in a random order throughout the network thus keeping the density throughout the network proportional. PEGASIS assumes to have global knowledge of the network topology, allowing it to use a greedy algorithm while constructing the chain. However, PEGASIS has a few drawbacks. First the clustering is based on random cluster heads. The chain described in PEGASIS may not be an optimal routing mechanism; other approaches such as directed diffusion [20] appear to give better performance. Again each node knowing the location information of all other nodes is tremendously costly in terms of memory and scalability.

The threshold-sensitive energy efficient sensor network (TEEN) [27] protocol uses a dynamic threshold calculation to decide the cluster heads on the fly. The nodes that are closer to each other, dynamically form a cluster, and the cluster heads broadcast two threshold values—one soft and one hard. Based on these two thresholds, the sensor nodes decide when to transmit a data packet, and accordingly reduce data delivery delay. The energy efficient clustering scheme (EECS) [33] for sensor networks selects cluster heads based on residual energy calculation. The hybrid energy efficient distributed clustering protocol (HEED) [34] incorporates additional topological information, like distance, degree, etc., to determine the optimal cluster heads.

All of the above clustering approaches use post-deployment sensor locations. However, multiple factors affect the clustering decision in sensor networks, leading to conflicting optimization criteria. Evolutionary multi-objective optimization methods have been well studied in the existing literature [1,2,10–12,21,22,31,35,36] to solve complex decision problems employing conflicting optimization criteria. Evolutionary multi-objective optimization methods use evolutionary algorithms to find out the best solution through multiple iterations, by traversing through all possible solutions. In [22], the authors have used an evolutionary multi-objective optimization method for joint sensor deployment and transmit power allocation in wireless sensor networks. In their proposed scheme, the authors have used two objective criteria: first to deploy the sensors such that minimum coverage issue gets solved, and second to find out the minimum transmit power for every sensor that ensures correct data decoding at the receiver. In [28], the authors have used evolutionary multi-objective optimization to find out energy efficient and loss aware data compression in sensor networks. Masazade et al. [29] have used the evolutionary optimization method to find out the threshold values for distributed detection of wireless sensor networks. In their works, the authors have studied the distributed detection problem in sensor networks and evaluated the sensor thresholds by formulating and solving a multi-objective optimization problem, where the objectives are to minimize the probability of error and the total energy consumption of the network. They have used a non-dominating sorting genetic algorithm method to find out the optimal threshold values. They have implemented their scheme in a simulator framework and observed that, instead of only minimizing the probability of error, multi-objective optimization provides a number of design alternatives that results in significant energy savings at the cost of slightly increasing the best achievable decision error probability.

7.3   CHPP: System Model and Multi-Objective Problem Formulation

As discussed earlier, this chapter assumes a heterogeneous sensor network architecture, where a set of low powered sensors are deployed in the target area to periodically sense data and forward it towards the cluster heads. The cluster heads are high power devices that accumulate the data from the sensors and forward it to the sink or the base station. Let $\mathcal{S}$ be the set of sensor nodes and $\mathcal{C}$ be the set of cluster heads. We assume that the sensors use a multi-hop forwarding mechanism to forward data to the cluster heads, such that the number of hops in the forwarding paths is bounded by k. This chapter proposes a mechanism for finding out the optimal positioning of the cluster heads subjected to the data forwarding and energy efficiency constraints.

For efficient data delivery to the cluster heads, the intermediate sensors use a breadth first search (BFS) forwarding tree rooted at the respective cluster heads. It can be noted that BFS-based data delivery is extensively studied in the existing literatures [4,7,8,24] that show the efficiency of bounded children BFS tree for low redundancy minimum delay data forwarding. In a BFS data forwarding tree, assuming the leaf nodes are at level zero, intermediate sensor nodes at level ℓ need to accumulate data from all its children in the rooted subtree and forward it to its parent. Further, to bound data aggregation complexity at intermediate sensor nodes, we assume that the number of children at every intermediate sensor is bounded by c. Then for an intermediate sensor node at level ℓ, the total number of sensor nodes in the rooted subtree of that node can be computed by the following recursion,

$$E(\ell) = c(E(\ell + 1) + 1) \tag{7.1}$$

where $E(\ell)$ is the number of children in the rooted subtree for a node at level ℓ. Solving Equation (7.1), $E(\ell)$ can be computed as

$$E(\ell) = \frac{c(c^{k-\ell}) - 1}{c - 1} \tag{7.2}$$

where k is the maximum height of the tree that depicts the maximum number of hops to forward data to the cluster head.

An intermediate sensor node at level ℓ needs to forward all the data traffic to its parent. Therefore, the energy consumption and delay in packet delivery of that node are proportional to $E(\ell)$. Therefore to reduce energy consumption and delay in packet delivery, $E(\ell)$ needs to be reduced by increasing number of cluster heads.

On the other hand, increasing the number of cluster heads arbitrarily reduces the network scalability and hierarchy of data forwarding. There needs to be some upper bound on the number of cluster heads in the network. The maximum number of cluster heads in the network is bounded by the minimum and maximum path length of a sensor node towards the cluster head. Let k_{min} and k_{max} be the minimum and maximum path length bound. The minimum path length ensures the minimum data aggregation level, and the maximum path length ensures bound on the data forwarding delay.

Let (x_i, y_i) denote the position of sensor node i. We define an indicator function,

$$\mathcal{I}_{x_i,y_i,x_j,y_j} = \begin{cases} 1 & \text{if } (x_i, y_i) \text{ is a parent node of } (x_j, y_j) \\ 0 & \text{Otherwise} \end{cases} \tag{7.3}$$

Then we define the distance objective $D(X,Y)$ as

$$D(X,Y) = \frac{\left[\sum\limits_{i \in \mathcal{C}} \sum\limits_{j \in \mathcal{S}} \mathcal{I}_{x_i,y_i,x_j,y_j} \right]}{k_{min}} - \frac{\left[\sum\limits_{i \in \mathcal{C}} \sum\limits_{j \in \mathcal{S}} \mathcal{I}_{x_i,y_i,x_j,y_j} \right]}{k_{max}} \tag{7.4}$$

It can be noted that $0 \le D(X,Y) \le 1$. Maximizing $D(X,Y)$ implies that the spread between the minimum and the maximum path length is maintained. $D(X,Y) = 1$ implies that both the minimum path length and maximum path length nodes are present in the network. The objective of this chapter is to find out the optimal positions for all sensors $(X = \{x_i\}, Y = \{y_i\})$ that minimizes $E(\ell)$ while maximizing $D(X,Y)$. It can be noted that these two objectives are conflicting objectives. Therefore, it is not possible to find out an optimal solution for this problem. Therefore, in this chapter we target to find out an *Pareto optimal* solution for this problem. Pareto optimality [36] is the state of allocation for resources where it is impossible to improve one objective better without making at least one other objective worse off.

7.4 CHPP: Evolutionary Multi-Objective Optimization and Pareto Optimality

Figure 7.1 shows a set of solution points for randomly deployed sensor nodes based on a uniform distribution scenario, with mean sensor distance 2 m and communication radius 5 m. k_{min} and k_{max} are chosen as 4 and 8, respectively. The different solution set of the CHPP can be spread in a two-dimensional space, with $D(X, Y)$ at the x-axis and $E(\ell)$ at the y-axis. The line on the figure indicates the set of Pareto optimal solutions for this problem.

As Figure 7.1 indicates there does not exist a unique solution of this problem that can maximize $D(X, Y)$ while minimizing $E(\ell)$. Therefore, in this chapter, we are interested in finding out a Pareto optimal solution of this problem. The concept of evolutionary multi-objective optimization [2, 10, 12, 36] is explored in this context to find a Pareto optimal solution. The goal of the CHPP is to give the network designers a diverse set of solution spaces based on the two conflicting objectives. In the following, we discuss the two extreme solution sets of the CHPP:

- Solution 1 (Optimal $E(\ell)$): This solution set favors the optimal value of the $E(\ell)$, while ignoring the effect of $D(X, Y)$. Such a solution increases the number of cluster heads in the network to reduce the average number of children per intermediate node. Such a solution decreases the data forwarding delay, however increases the average power consumption and

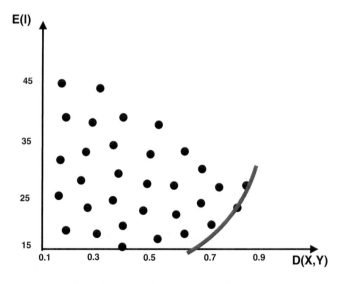

FIGURE 7.1: CHPP solution set and Pareto optimality

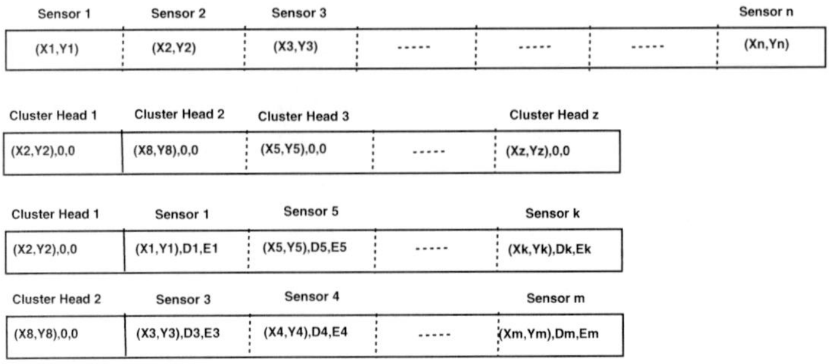

Sensor 1	Sensor 2	Sensor 3				Sensor n
(X1,Y1)	(X2,Y2)	(X3,Y3)	-----	-----	-----	(Xn,Yn)

Cluster Head 1	Cluster Head 2	Cluster Head 3		Cluster Head z
(X2,Y2),0,0	(X8,Y8),0,0	(X5,Y5),0,0	-----	(Xz,Yz),0,0

Cluster Head 1	Sensor 1	Sensor 5		Sensor k
(X2,Y2),0,0	(X1,Y1),D1,E1	(X5,Y5),D5,E5	-----	(Xk,Yk),Dk,Ek

Cluster Head 2	Sensor 3	Sensor 4		Sensor m
(X8,Y8),0,0	(X3,Y3),D3,E3	(X4,Y4),D4,E4	-----	(Xm,Ym),Dm,Em

FIGURE 7.2: Initialization of chromosomes for evolutionary optimization

cost of deploying the network, as cluster heads are special devices with extra configurations and device support.

- Solution 2 (Optimal $D(X, Y)$): Such a solution increases the spread of the cluster heads by bounding the maximum and minimum path lengths in the network. This solution decreases network power consumption and cost for deployment, however, increases data delivery delay, as aggregation level becomes high. With a long path length, some data experiences higher delivery delay that also increases the jitter in the network.

Therefore, in this chapter, we use the concept of evolutionary multi-objective optimization based on decomposition (MOEA/D) [35] to find out a set of solutions which are Pareto optimal, based on the trade-off between the above two extreme solution conditions. The MOEA/D has four steps:

1. Decompose the problem into a set of subproblems.

2. Initialize the solution sets.

3. Apply genetic operators to evolve the solutions to generate new solutions.

4. Check whether the stopping criterion is satisfied based on Pareto optimality.

7.4.1 Problem Decomposition

The first phase of MOEA/D is to decompose the vector optimization problem in a set of scalar subproblems. We use the weighted sum approach for scalar decomposition of the vector optimization problem. Let λ_i be the weight coefficient for the ith subproblem where $i1, 2, ..., m$ and m is total number of subproblems. Then the ith scalar optimization $\mathcal{S}(\ell, X, Y)$ is represented as follows:

$$\mathcal{S}(\ell, X, Y) = \lambda_i E(\ell) + (1 - \lambda_i) V(X, Y) \qquad (7.5)$$

Based on the positions of the sensors, ℓ can be represented in terms of sensor positions (X, Y). Therefore, the subproblem given in Equation (7.5) can be represented more generally as:

$$\mathcal{S}_i(X, Y) = \lambda_i E(X, Y) + (1 - \lambda_i)V(X, Y) \qquad (7.6)$$

Using a similar approach as proposed in [22], the scalar weight coefficients λ_i are calculated as:

$$\lambda_i = 1 - \frac{i}{m}$$

We initialize $\lambda_1 = 1$. The weight coefficients are used to decompose the vector optimization problem into a set of scalar optimization problems. This set of scalar optimization problems can be fed into the MOEA/D method to find out the set of Pareto optimal solutions. By varying the weight coefficients, we can find out different Pareto optimal solutions.

7.4.2   Initialization for Evolutionary Optimization

Next we need to define the solution set and its structure. Figure 7.2 shows the solution structure. In MOEA/D, solutions are represented as chromosomes. For CHPP, we define two fixed chromosomes and $|\mathcal{C}|$ number of dynamic chromosomes, where $|\mathcal{C}|$ denotes the number of cluster heads in the network. The first chromosome represents the position of the sensor nodes, and the second chromosome represents the position of the cluster heads. The dynamic chromosomes define the cluster, where the first entry is for the cluster head and the rest of the entries are for the sensors that belong to the cluster. Every entry has three tuples: the position of the node, the value of $E(\ell)$, and the value of $D(X, Y)$.

The initial positions of the sensors are generated randomly. We fix the number of cluster heads, $|\mathcal{C}|$, and select the cluster heads uniformly from the deployed sensors, so that every cluster contains almost equal number of sensor nodes. This is not an optimal solution, as the optimal values of $E(\ell)$ and $D(X, Y)$ depend on the position of the sensors. With the iterations based on genetic operators for MOEA/D, the proposed mechanism finds out the Pareto optimal positions of the sensors.

The initialization process generates a set of solutions based on the randomization method as discussed above. Let the initial solution set be $\mathbb{I}_{\text{CHPP}}$. This solution set is used for further operations for evolutionary optimization, as discussed in the next subsections.

7.4.3   Genetic Operations

In an evolutionary multi-objective optimization mechanism, genetic operators are applied over a solution to get a new solution. Following genetic operators are applied in sequence:

1. Selection operator

2. Crossover operator

3. Mutation operator

4. Repair operator

7.4.3.1 Selection Operator

The selection operator for evolutionary computing is responsible for the "Survival of the Fittest" concept, and it chooses the most promising solutions from the previous solutions set. According to evolutionary computing terminology, the previous solution set is termed as "parent" and the newly generated solution set by applying genetic operators are termed as "offspring." We use an M-tournament selection operator [21,35] that selects offspring solutions based on the measurement of Euclidean distance of their weights $(\lambda_1, \lambda_2, ..., \lambda_m)$. In this selection mechanism, two parent chromosomes are selected based on the closest Euclidean distance of their weights. These parent chromosomes are evaluated based on the optimization criteria $\mathcal{S}_i(X, Y)$, for the ith subproblem, as given in Equation (7.6). The best solutions are selected as offspring for further operations.

The functionality of the selection operator is given in Algorithm 3. The offspring solutions are then forwarded for the crossover operator.

Algorithm 3 Selection Operator

Input: Subproblem $\mathcal{S}_i(X, Y)$.
Output: Offspring solutions based on selection operator.

Select the solutions of M closest subproblems of $\mathcal{S}_i(X, Y)$ based on the measurement of Euclidean distance of the weights.
Evaluate the solution based on optimization criteria given in Equation (7.6).
Find the best solutions of the tournament.

7.4.3.2 Crossover Operator

The crossover operator combines two parent solutions $\mathcal{P}_1$ and $\mathcal{P}_2$ and generates a new offspring. A number of crossover operations are discussed in the existing literature [13–15, 32]. We use a window-based crossover operation, where a window size for solution i, denoted as W_i, is determined using the similar approach proposed in [22],

$$W_i = |\mathcal{S}| + |\mathcal{S}| \times (1 - \lambda_i) \tag{7.7}$$

where $|\mathcal{S}|$ is the number of sensors in the arena.

In the CHPP solution, crossover is done for all the chromosomes. Equation (7.7) gives the window size for the chromosome that determines the position of all the sensors. Let W_i^h and W_i^c denote the window size for the cluster head chromosome and the chromosomes corresponding to the individual cluster. These are calculated using a similar way as follows:

$$W_i^h = |\mathcal{C}| + |\mathcal{C}| \times (1 - \lambda_i) \tag{7.8}$$

$$W_i^c = |\mathcal{C}_i| + |\mathcal{C}_i| \times (1 - \lambda_i) \tag{7.9}$$

where $|\mathcal{C}_i|$ is the size of the ith cluster.

In the window-based crossover mechanism, two parent solutions $\mathcal{P}_1$ and $\mathcal{P}_2$ are selected based on the fitness criteria as discussed earlier, and a number of genes (entries in the chromosomes) are mutually swapped between the two parents to generate offspring solutions. The crossover procedure is shown in Algorithm 4.

Algorithm 4 Crossover operation

Input: Two parents $\mathcal{P}_1$ and $\mathcal{P}_2$.
Output: Offspring solutions after crossover.

Randomly generate an integer j from $\{1, 2, ..., W_i\}$.
Select j number of genes from $\mathcal{P}_1$ and swap the positions with the genes from $\mathcal{P}_2$, such that the genes are selected with an probability p_j.
Execute Step 1 and Step 2 for the cluster heads with W_i^h as the window size.
Based on the cluster heads and the sensors, define individual clusters.

7.4.3.3   Mutation Operator

After the crossover, the mutation operator is applied over the offspring solutions. The mutation operator selects genes randomly from the chromosomes, and alter the position of the sensors. It can be noted that the mutation operation is performed only on the set of sensors. The mutation operator changes the position of the sensors to explore the solution possibilities for different positions of the sensors. In the mutation operator, a sensor (gene) is randomly selected with probability p_m, and its position is shifted as:

$$X = X \pm f_r(x_d)$$

$$Y = Y \pm f_r(x_d)$$

where x_d is the minimum distance between two sensors and $f_r(x)$ is a function that selects a number randomly from $\{1, 2, ..., x\}$. It can be noted that in this scheme we only use integer positions of the sensors.

7.4.3.4   Repair Operator

The repair operator checks whether a solution generated by the crossover and mutation operators results in an offspring which is not valid. In our scheme, a solution is called invalid if one of the following cases occurs:

1. Two sensors have the same location.

2. Two cluster heads are very close such that the distance between them is less than the minimum threshold.

3. The number of clusters in a cluster head is below a threshold.

If one of these cases occur, the repair operator re-performs the crossover and the mutation operations to generate a different offspring. These newly generated offspring solutions are used in further iterations and the previous solutions are discarded.

7.4.4   Termination Condition

The termination condition determines when the iteration stops and a solution is obtained. In our solution approach, the best solution found until now, based on the optimization criteria given in Equation (7.6), is stored. We set a parameter g_m that defines the maximum number of generations for the MOEA/D process. Once the number of iteration reaches g_m, the proposed scheme terminates, and returns the best solution obtained so far.

7.5   Performance Evaluation

The proposed scheme is implemented in an NS-3.19 [19] network simulator framework. An arena of $1000 \times 1000 \ m^2$ is selected with a maximum of 200 number of sensors and 10 cluster heads. The communication radius is taken as $5m$ at minimum and $10m$ at maximum and k_{min} and k_{max} are set to 5 and 20, respectively. Channel bandwidth is considered as 2 Mbps. The setups for the parameters of MOEA/D are given in Table 7.1.

First we show the performance of the CHPP for minimal delay data forwarding and average power consumption. We have compared the performance of CHPP with other three clustering and sensor deployment protocols: TEEN [27], HEED [34], and EECS [33]. The TEEN protocol selects the cluster heads dynamically through an energy threshold observation. The HEED protocol extends the basic LEACH clustering protocol with the residual energy and topology parameters for cluster head selection metric. The EECS protocol selects the cluster head with more residual energy. It can be noted that all these protocols use dynamic cluster head selection through residual

TABLE 7.1: Parameter Settings for Evolutionary Optimization

Parameter	Max Value	Min Value
Crossover probability	0.1	0.8
Mutation probability	0.05	0.3
Maximum generation size	100	300
Population size	50	200
Tournament size	50	200
Neighborhood size	30	100

TABLE 7.2: Average Data Forwarding Delay (ms)

#Sensors	#CH	TEEN	EECS	HEED	CHPP
50	5	21.2	20.1	18.2	14.1
75	8	23.5	22.4	20.8	16.3
100	10	28.9	26.7	24.2	19.8
150	12	32.2	31.8	29.7	23.7
200	15	42.1	39.4	37.2	32.1

TABLE 7.3: Average Energy Consumption (Watt-Hour)

#Sensors	#CH	TEEN	EECS	HEED	CHPP
50	5	121.2	112.3	102.5	98.2
75	8	112.3	101.3	91.7	89.3
100	10	110.7	96.5	88.5	86.5
150	12	109.8	95.3	86.5	83.7
200	15	108.6	94.1	85.4	81.6

energy calculation. However, the proposed scheme uses a static cluster head selection that minimizes the path length and data aggregation level which in turn minimizes the energy consumption. Further, it can be noted that in the existing schemes, sensor nodes are considered as static and post-deployment cluster head selection, where CHPP considers sensor positioning and cluster head selection simultaneously with an objective to minimize data forwarding delay.

Table 7.2 compares the average data forwarding delay for different schemes. #Sensors denotes the total number of sensors deployed and #CH denotes the total number of cluster heads. In the simulation scenario, all sensor nodes generate constant bit rate (CBR) traffic with a data generation rate of 64 Kbps and forward the data to the cluster heads in a multi-path forwarding mechanism. The data forwarding delay is the time between when the data packet is scheduled in the interface queue for the originating sensors and the time when the data packet is delivered. The table indicates that the proposed optimization method significantly reduces the data forwarding delay.

TABLE 7.4: Convergence and Optimality Ratio Analysis

#Sensors	#CH	Iterations to Converge	Optimality Ratio
50	5	30	1.8%
75	8	42	2%
100	10	54	2.2%
150	12	62	2.7%
200	15	78	3.2%

Table 7.3 compares the average energy consumption measured in terms of Watt-hour. We use the energy configuration profiles for Micaz sensor motes [23]. The table indicates that the proposed scheme also reduces energy conservation. It can also be observed that as the number of sensor and the number of cluster heads grow, average energy consumption decreases. The proposed scheme maintains a bound on the number of children for every sensor, that in turn reduces energy loss due to contention during data delivery.

Next we evaluate the convergence speed for MOEA/D and compute the optimality ratio that gives the percentage difference between the nearest Pareto optimal solution and the solution obtained through the proposed approach. Table 7.4 summarizes the results. The table shows that even with a large number of sensor nodes, the number of iterations required to find out the best solution is less than 100, and the optimality ratio is less than 5%. This indicates that the solution obtained through MOEA/D is not worse than 5% of one of the Pareto optimal solution. It can be noted that, in the simulation setup, every scenario is executed for 10 times and the average result is reported in this chapter.

7.6 Conclusion

This paper proposes a cluster positioning problem for a heterogeneous sensor network, where both the position of the cluster heads and the sensor nodes need to find out simultaneously. We have used the evolutionary multi-objective optimization method based on decomposition to solve this problem and to find out the optimal deployment scenario for a given set of constraints, number of sensors, number of cluster heads, and other communication parameters of the sensor nodes. The performance of the proposed scheme is analyzed through simulation results, and compared with other related schemes proposed in the literature.

Bibliography

[1] H. A. Abbass, R. Sarker, and C. Newton. PDE: a pareto-frontier differential evolution approach for multi-objective optimization problems. In *Proceedings of the 2001 Congress on Evolutionary Computation*, volume 2, pages 971–978, 2001.

[2] A. Abraham and L. Jain. *Evolutionary multiobjective optimization*. Springer, 2005.

[3] J. Albath, M. Thakur, and S. Madria. Energy constraint clustering algorithms for wireless sensor networks. *Ad Hoc Networks*, 11(8):2512–2525, 2013.

[4] K. Arisha, M. Youssef, and M. Younis. Energy-aware TDMA-based MAC for sensor networks. In *System-level power optimization for wireless multimedia communication*, pages 21–40. Springer, 2002.

[5] S. Bandyopadhyay and E. J. Coyle. An energy efficient hierarchical clustering algorithm for wireless sensor networks. In *Proceedings of the Twenty-Second Annual Joint Conference of the IEEE Computer and Communications*, volume 3, pages 1713–1723, 2003.

[6] M. Chatterjee, S. K. Das, and D. Turgut. WCA: A weighted clustering algorithm for mobile ad hoc networks. *Cluster Computing*, 5(2):193–204, April 2002.

[7] S. Chen, M. Huang, S. Tang, and Y. Wang. Capacity of data collection in arbitrary wireless sensor networks. *IEEE Transactions on Parallel and Distributed Systems*, 23(1):52–60, 2012.

[8] T.-Shi Chen, H.-W. Tsai, and C.-P. Chu. Gathering-load-balanced tree protocol for wireless sensor networks. In *Proceedings of the IEEE International Conference on Sensor Networks, Ubiquitous, and Trustworthy Computing*, volume 2, pages 8–13. IEEE, 2006.

[9] X. Chen, Z. Dai, W. Li, Y. Hu, J. Wu, H. Shi, and S. Lu. ProHet: A probabilistic routing protocol with assured delivery rate in wireless heterogeneous sensor networks. *IEEE Transactions on Wireless Communications*, 12(4):1524–1531, April 2013.

[10] C. A. C. Coello. Recent trends in evolutionary multiobjective optimization. In *Evolutionary Multiobjective Optimization*, pages 7–32. Springer, 2005.

[11] C. A. C. Coello. Evolutionary multi-objective optimization: a historical view of the field. *IEEE Computational Intelligence Magazine*, 1(1):28–36, 2006.

[12] K. Deb, L. Thiele, M. Laumanns, and E. Zitzler. *Scalable test problems for evolutionary multiobjective optimization*. Springer, 2005.

[13] K. Deep and M. Thakur. A new crossover operator for real coded genetic algorithms. *Applied Mathematics and Computation*, 188(1):895–911, 2007.

[14] A. E. Eiben, R. Hinterding, and Z. Michalewicz. Parameter control in evolutionary algorithms. *Evolutionary Computation, IEEE Transactions on*, 3(2):124–141, 1999.

[15] C. Emmanouilidis, A. Hunter, and J. MacIntyre. A multiobjective evolutionary setting for feature selection and a commonality-based crossover operator. In *Proceedings of the 2000 Congress on Evolutionary Computation*, volume 1, pages 309–316, 2000.

[16] M.J. Handy, M. Haase, and D. Timmermann. Low energy adaptive clustering hierarchy with deterministic cluster-head selection. In *Mobile and Wireless Communications Network, 2002. 4th International Workshop on*, pages 368–372. IEEE, 2002.

[17] M. S. Hefeida, T. Canli, and A. Khokhar. CL-MAC: A cross-layer mac protocol for heterogeneous wireless sensor networks. *Ad Hoc Networks*, 11(1):213–225, 2013.

[18] W.B. Heinzelman, A.P. Chandrakasan, and H. Balakrishnan. An application-specific protocol architecture for wireless microsensor networks. *IEEE Transactions on Wireless Communications*, 1(4):660–670, Oct 2002.

[19] T. R Henderson, M. Lacage, G. F. Riley, C. Dowell, and J.B. Kopena. Network simulations with the NS-3 simulator. *SIGCOMM demonstration*, 2008.

[20] C. Intanagonwiwat, R. Govindan, and D. Estrin. Directed diffusion: a scalable and robust communication paradigm for sensor networks. In *Proceedings of the 6th annual international conference on Mobile computing and networking*, pages 56–67, 2000.

[21] H. Ishibuchi, Y. Sakane, N. Tsukamoto, and Y. Nojima. Adaptation of scalarizing functions in MOEA/D: An adaptive scalarizing function-based multiobjective evolutionary algorithm. In M. Ehrgott, C. M. Fonseca, X. Gandibleux, J.-K. Hao, and M. Sevaux, editors, *Evolutionary Multi-Criterion Optimization*, volume 5467 of *Lecture Notes in Computer Science*, pages 438–452. Springer Berlin Heidelberg, 2009.

[22] A. Konstantinidis, K. Yang, Q. Zhang, and D. Z.-Yazti. A multi-objective evolutionary algorithm for the deployment and power assignment problem in wireless sensor networks. *Computer Networks*, 54(6):960–976, April 2010.

[23] M. Krämer and A. Geraldy. Energy measurements for MICAZ node. *5. GI/ITG KuVS Fachgespräch Drahtlose Sensornetze*, 2006.

[24] M. Li, Y. Wang, and Y. Wang. Complexity of data collection, aggregation, and selection for wireless sensor networks. *IEEE Transactions on Computers*, 60(3):386–399, 2011.

[25] S. Lindsey and C. S. Raghavendra. PEGASIS: Power-efficient gathering in sensor information systems. In *IEEE Aerospace conference proceedings*, volume 3, pages 3–1125, 2002.

[26] V. Loscri, G. Morabito, and S. Marano. A two-levels hierarchy for low-energy adaptive clustering hierarchy (TL-LEACH). In *IEEE Vehicular Technology Conference*, volume 62, page 1809, 2005.

[27] A. Manjeshwar and D. P. Agrawal. TEEN: a routing protocol for enhanced efficiency in wireless sensor networks. In *Parallel and Distributed Processing Symposium, International*, volume 3, pages 30189a–30189a. IEEE Computer Society, 2001.

[28] F. Marcelloni and M. Vecchio. Enabling energy-efficient and lossy-aware data compression in wireless sensor networks by multi-objective evolutionary optimization. *Information Sciences*, 180(10):1924–1941, 2010.

[29] E. Masazade, R. Rajagopalan, P. K. Varshney, C. K. Mohan, Gullu Kiziltas Sendur, and Mehmet Keskinoz. A multiobjective optimization approach to obtain decision thresholds for distributed detection in wireless sensor networks. *IEEE Transactions on Systems, Man, and Cybernetics, Part B: Cybernetics*, 40(2):444–457, 2010.

[30] M. A. Razzaque, C. Bleakley, and S. Dobson. Compression in wireless sensor networks: A survey and comparative evaluation. *ACM Transactions on Sensor Networks*, 10(1):5:1–5:44, December 2013.

[31] K. Sindhya, K. Miettinen, and K. Deb. A hybrid framework for evolutionary multi-objective optimization. *IEEE Transactions on Evolutionary Computation*, 17(4):495–511, 2013.

[32] W. M. Spears. Adapting crossover in evolutionary algorithms. In *Evolutionary programming*, pages 367–384, 1995.

[33] M. Ye, C. Li, G. Chen, and J. Wu. EECS: an energy efficient clustering scheme in wireless sensor networks. In *24th IEEE International Performance, Computing, and Communications Conference*, pages 535–540. IEEE, 2005.

[34] O. Younis and S. Fahmy. HEED: a hybrid, energy-efficient, distributed clustering approach for ad hoc sensor networks. *IEEE Transactions on Mobile Computing*, 3(4):366–379, 2004.

[35] Q. Zhang and H. Li. MOEA/D: A multiobjective evolutionary algorithm based on decomposition. *Evolutionary Computation, IEEE Transactions on*, 11(6):712–731, Dec 2007.

[36] E. Zitzler and L. Thiele. Multiobjective evolutionary algorithms: a comparative case study and the strength Pareto approach. *IEEE Transactions on Evolutionary Computation*, 3(4):257–271, 1999.

Part III

Advanced Topics

Chapter 8

Ubiquitous Context Aware Services

P. Venkata Krishna

Sri Padmavati Mahila University

V. Saritha

VIT University

S. Sivanesan

VIT University

8.1 Introduction

The increased advancement in communication and computing technologies has opened many avenues in services industries that support many stakeholders like professionals, farmers, various governmental agencies and private agencies. These technological developments increasingly affect all aspects of work programs like operations, administration, and management. The huge amount of information is available due to advanced data handling technologies. The biggest question is whether the right information is available to the appropriate user at the opportune time. Instead of providing answers, this question has generated many more questions like what makes the right information, who is the appropriate user, and when will there be an appropriate time? Within any context, the right information could be delivered to appropriate users at an opportune time. Hence, context awareness needs to be built into every system with minimal disturbance to the existing technologies and platform.

This chapter focuses on how wireless sensor networks are used in building context aware services with appropriate illustrative examples. It presents ubiquitous context aware middleware architecture for precision agriculture to solve major issues such as waste of water, improper application of fertilizer, choice of wrong crops and season, poor yield, lack of marketing, and other issues. This chapter also focuses on how wireless sensor networks are used in building context aware services with appropriate illustrative examples.

In recent times wireless sensor networks (WSNs) have gained significance in the field of application development due to extensive usage of sensor devices to monitor and analyze the information about the environment such as earthquake detection and agriculture. The capability of WSNs for detecting environmental changes, called context changes, by means of sensors, made WSNs more popular even in the daily tasks of humans.

Because the sensor nodes are heterogeneous, the information carried by them is complex. For example, in the context of patients, information is derived from day-to-day health monitoring, frequency, and intensity of movement. Grouping of contexts allows communication heterogeneous sensor devices to represent the same information in different formats. Even though the WSNs are very capable in handling information on heterogeneous sensor devices, significant research must be done in the field to gather the information.

The context aware service is a computing technology that integrates data related to the area of a mobile user to deliver added significant services. The real world characteristics such as temperature, time, and location, can be referred to as context. With the availability of ICT such as cloud computing, heterogeneous networking, crowd sourcing, web services, and data mining, sharing of the information at any time anywhere has become possible. However, this will bring out lot of challenges such as incompatibility in standards, data portability, data aggregation, data dissemination, and differential context communication overhead.

Even though the ICT has changed many aspects of human lifestyle, work places, and living places, there is lot of gap in the agriculture sector for ICT to be of value. This creates a digital divide between farming and other industries. Farming consumes large amounts of natural resources such as water, energy, and fertilizers that could escalate global warming, soil degradation, and depletion of ground water. By integrating the needs of the farming industry with relevant technologies, global warming can be considerably reduced.

Interaction is the term for the next decade. The interactions between people, businesses, devices, and technologies will create more novel challenges and opportunities. The interaction among people, businesses, devices, and technologies will create more novel prospects and challenges. Interaction is very important especially among the stakeholders such as the public, market, government, and other agencies related directly or indirectly with the farming communities.

Interactions play a major role in creating context aware precision agriculture. Neural networks (NNs) represent a biological approach for retrieving information from WSNs. NNs may be classified as artificial neural networks (ANNs), Bayesian networks (BNs), and hidden Markov models. This chapter explains context aware sensing based on these three models and addresses some of the issues raised by context aware sensing models.

8.2 Context Awareness in WSNs

WSNs have been used in various applications such as body sensor networks (BSNs), earth sensing, and environmental sensing. For example, body sensor networks are used to monitor patient data such as heart rate, blood pressure, oxygen condition, and other parameters. To monitor the human body requires a BSN framework.

Various context aware algorithms with different characteristics for performing certain functions can be used to retrieve information. The basis of context aware sensing is to gather information from all sensor nodes in a lower hierarchy; the gathered information services as input in the next context.

According to [8], it is common to order input data in a certain way through appropriate clustering. The data is clustered in sub groups. The distance among data entries in the same clusters are small and the distance among data entries of different clusters are large. Cluster inputs must be associated with context to allow information to be retrieved from sensor nodes. The classification approach for sensor networks assigns input vectors to a context through user labels.

WSNs are commonly used in various applications such as health care, smart homes, construction, and many more. All these environments have unique contexts and require specific frameworks to handle their different needs. The diversity of needs for context aware services and dynamic nature of WSNs create design challenges. The key issues to be considered in WSNs are node failure, noise interference, integration of multisensor data, smooth context recognition, long-term use of applications, number of nodes in network, and selection of appropriate features.

A context aware system requires procurement, depiction, storage, analysis, and revision modules. The procurement model senses and collects contexts.

Sensors receive huge amounts of information that must be depicted in a standard format so that it can be reused as required. The depiction module formats and stores data received from the sensors. The stored data is analyzed and converted into the required form by the analysis module.

After analysis is complete, the information is used to provide client services. A suitable service is chosen from those available based on the interpretation of context information, after which appropriate information is delivered to the client. The behaviours of the service and the device are adapted according to the context [16].

According to [8], context aware computing is a model for software engineering capable of managing all aspects of building a system to cover all parameters of a context such as identity, time, location, and activity by adapting to the conditions existing at runtime [22]. Current programming environments do not allow context aware mechanism to meet needs and researchers should focus on developing efficient context oriented mechanisms [4].

Information and communication technologies have been used to automate farming tasks such as irrigation, planting, and harvesting, and even for supervision and management. Zhang [25] developed a network using WSNs to observe plant nursery conditions such as soil, and air moisture, humidity, temparature, and light intensity. One of the core aspects of farming is water use management. Several proposals suggested designs of sensor networks for monitoring water supplies for plants and ensuring optimum utilization [1] [3] [9] [11] [15] [19].

Tapankumar et al. [3] developed an architecture for and built a drip irrigation management system using a computer system with the ability to acquire information from distant locations. Kim et al. [11] designed an irrigation system capable of monitoring the soil moisture, temperature, and sprinkler positions using electronically managed sensor devices. All these parameters could be managed from a remote location through technologies such as Bluetooth and GPW.

Fungus infections and pests are prime concerns of the farming community. A sensor-based network has been proposed to address these concerns. Baggio [2] developed a design to combat infections in potato crops. Sensors measured humidity and temperature. Analysis of fluctuations allowed farmers to reduce the overall impacts of infections.

Cattle forming requires green fields for grazing and thus has a close tie to the farming community. Wark et al. [24] at Australia's **CSIRO ICT** Center designed a self-managing sensor node capable of monitoring the activities of cattle and assess and control pastures. Photographic sensing devices were used to study grass growth and this information meant the animals could move to greener pastures as required. Animal habits such as grazing, sleeping, and ruminating were also observed.

8.3 Architecture for Precision Agriculture

The farming lifecycle is highly complex, dynamic, and heavily influenced by environmental constituents such as soil, weather, market, seasons, living space, and lifestyle. Each of these constituents might have direct or indirect disruptive influence on the others. There is no single solution for all issues in farming. Hence, there is a strong need for a solution which is adaptable and feasible for the farming sector. Here, we present a flexible and viable solution discussed by Krishna et al. [18]. They defined context aware middleware (CAM) that consists of environment, ambience and collection of states. The following modules were described:

- Ambience monitoring agents (AMA): The CAM learns its context from environment through AMA. The AMAs play a major role in sensing various physical phenomena such as temperature, rainfall, light, wind speed, wind direction, moisture, and snow fall through relevant sensor devices.

- Context assemble (CA): A sub-system that assembles different contexts based on their event-state relationship.

- Context classifier (CC): A sub-system based on context sequences. CC assigns priority for each context sequence.

- Advanced context modeler (ACM): This builds refined context repositories that could be used in end-user applications. Look-up tables are created by ACM for every context with relevant action sequence.

- CStore: The ACM helps context store (CStore) learn all the contexts that could be used by the farmer console to take actions based on context.

Precision farming is developed using CAM based on pre-farming, mid-farming and post-farming. The pre-farming involves off-field activities such as collection of inputs such as seasonal facts, market demands, and suitable crops. It also includes on-field activities such as nursery production, main field preparation, and initiation of data acquisition [18]. Tasks involved in mid-farming are on-field activities such as crop transplanting, irrigation, pest control, infection control, and fertilizer utilization management. The post-farming involves harvesting, yield monitoring, yield grade processing, conditioning, packaging, demand identification, and supply. A successful farmer must carry out these sub-tasks. Precision farming is the most viable solution for all the farmers but achieving it is challenging. The basic building blocks of precision farming include [18]:

Layer 1 is represented by various stakeholders such as farmers, the public, e-commerce industries, data centers, mail facilities, and web interfaces. Farmers represent the farming community; the public represents people other than the farmers. The on-line trading agencies and virtual malls such as Amazon and eBay are represented as e-commerce industries. The governmental agencies such as meteorological centers, national information centers, and private weather monitoring stations are represented as data centers. The social networking sites such as Facebook, Twitter, BlogSpot, Google, and Yahoo are symbolized as mail facilities and web interfaces.

Layer 2 consists of a few core elements such as smart software agents, context aware middleware, and a data acquisition system. At this layer, smart software agents do the vital job of extracting the on-line information from stakeholders in layer 1. These agents constantly listen to the stakeholders at layer 1 and mine the information based on the subject of interest. For further processing, these agents also act as adapters and convert the information into common formats.

Figure 8.1 depicts the operation of CAM with a simplified illustration of workflow. The software agents select an important event from a known third party resource. For example, an agent has received data on a flooding event in Madagascar and another agent has received cyclone event data in Cuba. The captured events will be sent to context aware middleware. The CAM analyzes these events and frames a context that explains the destruction of banana and vanilla plants in Cuba and Madagascar, respectively.

The CAM creates new context which identifies the demand for banana and vanilla crops and also finds viable areas for growing banana and vanilla crops through advanced context modeler modules in the CAM. In this sample workflow, CAM identified India as a potential place for growing banana crops and Europe as a place for growing vanilla. This contextual information will be sent to farmers who already registered with the CAM-based network.

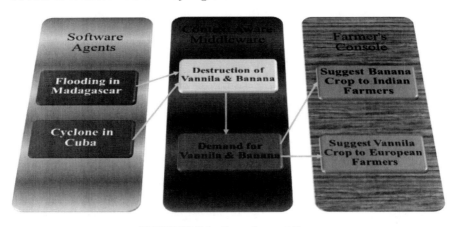

FIGURE 8.1: Sample workflow

Figure 8.2 provides better understanding about how contextual information is classified for a crop. This table captures important contextual information for the farming lifecycle of crops.

Crops		Soil		Seasons		Weather		Markets	
Banana	**Paddy**	**Clay**	**Sandy**	**Winter**	**Summer**	Rain	Snow	**Direct**	**Indirect**
Sprouting	Germination	Small Porosity	Large Porosity	Cold Waves	Heat Waves	Low	Low	Interactive	Non-interactive
Leaf development	Leaf development	Large surface area	Small surface area	Short day Long night	Long day short night	Medium	Medium	Low overhead	High overhead
Sucker formation	Tillering	Low permeability	High permeability	Snow or Cold rain	Forest fire	Heavy	Heavy	Cheaper	Costlier
Flowering	Panicle formation	Poor water recharge	Good water recharge	Very cold	Drought	Flooding	Flooding	Single channel	Multi-channel

FIGURE 8.2: Sample contextual information of farming lifecycle [18]

Figure 8.2 highlights a portion of contextual data which belongs to stakeholders in CAM. For example, if clay is considered as a main context, subcontexts such as small porosity, large surface area, low permeability and poor water recharging ability will allow better contextual mapping with other stakeholders. CAM helps to frame larger contexts such as "Field with clay soil is susceptible to water-logging which retards the growth of banana" and "Field with clay soil susceptible to water logging may not affect rice cultivation to larger extent." Hence, CAM creates flexible and feasible solution precision agriculture by providing context aware solutions.

Figure 8.3 illustrates the analysis of performance using context awareness in precision farming. In general, there is an assumption that paddy consumes heavy amounts of water, but in the trial run CAM consumed close to 40% less than the manual mode of cultivation. This is possible because irrigation will be done by CAM purely based on multiple fine grained contexts. The irrigation is not done by static timer based systems. The CAM analyzes the paddy lifecycle, type of soil, and type of weather and then creates optimum contexts for irrigation.

The next criterion reveals the amount of fertilizer used to cultivate the paddy in both manual and CAM modes. There is a savings of 17% in fertilizer by CAM based cultivation over manual farming. In manual mode, farmers use fertilizers based on their experience. Dynamic changes in season, soil, weather, and unplanned paddy impact farmers strategies. Due to the context awareness of the paddy lifecycle, CAM identifies the optimum requirement for fertilizers at various growth stages of paddy and reduces fertilizer consumption.

The third criterion reveals how effectively pests can be controlled. Again, CAM provides better results over manual methods by almost 35%. The effectiveness of pest control largely depends on factors such as how quickly effects

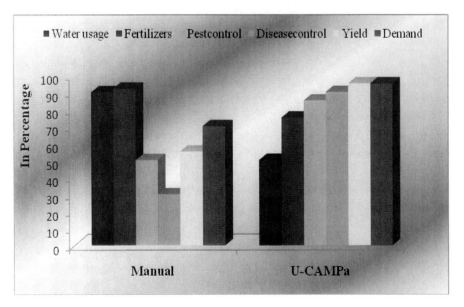

FIGURE 8.3: Performance analysis for manual and CAM-based paddy culti-vation

may be identified, prior history of paddy variety, weather, and seasons. Since, CAM thrives on all these factors pest control may be effective where manual modes may not.

Another criterion is how to control infections optimally. It depends on factors such as previous crop history, nutrient deficiencies, paddy varieties, soil conditions, weather, and season. Manual methods do not consider all these factors collectively and thus fall short by almost 60% in comparison with CAM. The fifth criterion concerns total yield percentage of paddy. The improved context afforded by CAM allowing better selection of paddy variety, soil, season, optimal use of water and fertilizers, effective pest control, and optimal disease control strategies helped CAM achieves a 40% increase in paddy yield over manual methods.

The final criterion predicts how effectively paddy may be sold in the mar-ket. Most farmers using manual methods have no information about market demand when they attempt to choose the best paddy variety. They are often disappointed by lower prices for non-premium paddy. CAM creates a context that allows better choices of varieties that can command higher market prices.

8.4 Context Extraction Using Bayesian Belief Networks

8.4.1 Bayesian Belief Networks (BBN)

A Bayesian belief network can be represented as a directed acyclic graph (DAG) with a set of nodes and edges. Each node of the graph denotes an arbitrary variable and each arc denotes a probable dependency directly between two variables. The local distributions of each node are multiplied by the distributions of its parents to calculate a joint probability distribution conveyed by a BBN. The dependencies among the nodes are depicted using DAG which helps in providing the qualitative function of a BBN. Some variables in the network may not have predecessors. For such nodes, earlier probability distributions are required during quantification. The nodes which have predecessors can proceed with conditional probability distributions [5].

8.4.2 Online Data Cleaning for Context Extraction

In cases where the server and client are in different layers, context extraction architecture is required. In WSNs, the nodes are at a layer different from the layer where the base station is run in [18]. The steps involved in processing the information at sensor nodes are:

- Nodes sense raw data.

- Raw sensor values are quantized based on threshold value.

- Node and cluster identifications and time data are inserted.

As the performance of BBN on raw data may not be better, quantization is required. The sampling instructions are transmitted to the base station. The steps involved in processing the information at base station are:

- Nodes are clustered.

- BBN is constructed.

- Data cleaning is carried out. If an outlier is introduced or data is missing, then BBN identifies and clarifies them.

- Rule-based mapping is carried out depending on the context semantics defined. Rule-based mapping maps the context features to abstract context situations [6].

During the transmission or because of conceded sensor nodes, errors may occur. Rule-based mapping is used to deliver the consistent and energy efficient context.

Outliers are introduced into the data because of faulty sensor nodes. BBN is used as a classifier to detect outliers. Classification involves classifying a class variable C from attribute variables. The probabilistic inference is calculated and utilized in approximating the probability of a group of query nodes provided with some sort of confirmation. In BBN, this process is referred as belief propagation. The variation in the probability of actual data and expected data is estimated using belief propagation. The information is said to have outliers if the variation level is high [20].

The values missed due to reasons like node failure, bad weather conditions, packet loss, congestion, and collision may be recovered using error detection and correction methods or by adding extra bits before transmission, but this kind of recovery is helpful only when the missing frequency is low. Hence, BBN is used to make recovery efficient when the frequency of missing is high. BBN is a soft computing technique where learning takes place. The learning process is carried out by using only correct data. Data with missing values are rejected. Then BBN is utilized to deduce the class of missing value with the help of the inference algorithm to project the missing value with its classifying behavior. There is a chance of missing the values from the data sensed by various sensors. Even then the BBN is capable of deducing the missing values.

8.5   Self-Organizing Map-Based Context Identification

8.5.1   Unsupervised, Dynamic Identification of Physiological and Activity Context in Wearable Computing

Wearable computing is used to sense the client and its present condition. By approximating physical status-like level of pressure, motion, movement outline, and ambient context data, the present condition can be estimated. Unsupervised learning is possible since the vivid label is not required for the context to be utilized to produce an adaptive and contextually delicate response. The recognition of the context without the need of external observation is called unsupervised learning [14].

8.5.2   Offline Algorithm

Initially, the information is preprocessed. The sensor values make the Kohonen self-organizing map easy to learn. The map codebook vectors are clustered using the standard k-means clustering algorithm. In the process of clustering, the value of k is selected [7]. Then these cluster changeover probabilities are used in the learning of the first order Markov model on which the graph reduction technique is made functional. All the temporary or short-lived states are removed in this graph reduction phase. The k-means algorithm is used to reallocate the code vectors of removed states to the permanent states. The final clusters are utilized in the classification of test samples. The first order Markov model on the final clustering can assist estimation.

As the independent component analysis (ICA) is not computationally cost effective, principle component analysis (PCA) is used for online algorithms on wearable computers to reduce the dimensions. The dimensions are further reduced by using Kohonen self organizing maps. The input data is projected onto the first five main elements as the top five eigenspaces are mostly used to describe the sensor variance cases of high and low data rate. Batch training algorithms [12] are used in machine learning.

The learning process of a self organizing map completes the process of clustering. In order to achieve the typical size of cluster, the codebook vectors are clustered together. The clusters are identified and the structures are developed using a U-matrix [23]. The evolving clusters have borders that are very sharp. For this reason, the utilization of a U-matrix alone makes the process tedious. Therefore, codebook vectors are clustered using k-means clustering also. No labeling is used in determining the clusters. Initially large k-values are used for clustering.

8.5.3   Online Algorithm

The clusters signifying the context alterations are predicted to be identified as the information preprocessing as anticipated to level the unexpected modifications of the context. The training performance of the system is highly impacted by these kinds of unexpected modifications as the system must depend on vigorous context state approximations as the beginning of training practice patterns. To detect temporary states, changeover probabilities are considered. The cluster changeover probabilities are used in training process of the first order Markov model. The probability of not being in the same state for a long time is referred as low loop probability. This low loop probability is used to categorize the temporary states.

The states that are categorized as temporary are removed by using a graph reduction algorithm on the Markov model. The codebook vectors which are not clustered are assigned to the new clusters after the removal of the temporary clusters with the help of the k-means clustering algorithm. The minimized set of clusters is used to train the new transition probability Markov model which is used to estimate the probable changes in the context.

The previous training data is compressed and new data is integrated to form a new training sample used to improve the model. The compression of the previous training data is carried out by selecting the data at regular intervals of space and eliminating the remaining data. The process of developing a Markov model, graph reduction, and another Markov model, and another graph reduction process is continuous. This continuous process indicates machine learning.

By including these arbitrarily modified cluster centers, possible new contexts can be recognized. The sensor node properties and the positions of the sensors are deliberated as the directions of the main module. Hence, it is assumed that the primary approximation of the main module direction is determined by the information obtained from a restricted time interval. New information in real time can be anticipated to the earlier determined main modules by the direct change determined by these directions. Once the variation is below the predetermined threshold, recalibration can be activated.

8.6 Neural Network Model for Context Classification in Mobile Sensor Applications

In this section, the concentration is on how an application helps monitor the health condition of a patient. For mobile sensor applications, the focus is on the architecture of the model.

The architecture uses a mobile phone carried by the user to communicate with sensor environments. A wireless router attached to the phone communicates with other sensor devices in an ad hoc manner. The exact communication model is not required to manage the user locations. The CONSORTS-S [21] is a mobile sensing platform to validate the mobile sensing application, popularly used in Japan. Health care services as an application for mobile users by accessing surrounding context aware services are implemented in this platform.

8.6.1 Context Recognition in Wireless Sensor Networks

Context aware services use the neural network model to identify, build, and classify various contexts appropriate to mobile sensor applications. The neural network model for mobile sensor applications consists of three layers: input, hidden, and output. The input layer is responsible for collecting the information from the wearable sensors and environmental sensors. The hidden layer consists of router, mobile device, and sensor middleware. The router collects the information and forwards it to the mobile phone. The phone sends this information to the sensor middleware. The sensor middleware is responsible for fusion and classification of data. Finally the processed data is stored in the database.

8.6.2 CONSORTS-S: Mobile Sensing Platform

Consider the architecture of the CONSORT-S model [21]. It contains two additional components: mobile sensor router and sensor middleware. For communication, the mobile is linked with a mobile sensor router. Sensor middleware on a remote server performs the fusion and analysis of the collected data from the sensor nodes.

8.6.3 Mobile Health Care Services

The health care services in relation to context aware services in WSNs are considered. The mobile health care services are used to maintain and promote health of patients by monitoring both biological information and environmental information, within a context collected from both conditions. Different architectures are considered.

The health care model monitors the health of the patient by regulating the room temperature and collecting the context from wireless sensor nodes through a network dedicated to the communication [17]. A sensor dedicated to collecting the physical signals must be attached to the user's chest, and monitors the motion of the user. To monitor temperature a sensor is placed in a room. Temperature sensors are placed in all rooms where the user will stay. The user will recognize the room temperature automatically by communicating with the sensor devices. The sensed context is transferred to the mobile phone through the mobile sensor router as explained previously.

In this section, an example for context aware services to collect the information from wireless sensor networks is defined. The mobile sensor application is considered along with the different architecture by using neural networks. The prototype model provides health care services by communicating with mobile phones [10] and with the surrounding wireless sensor networks. In these sensor networks, the context is collected from the surrounding sensor nodes [13] and they can publish sensing information anywhere at any time.

8.7 Conclusion

The expansive progress of information and communication technology has altered many facets of human lifestyle, work places and living spaces. Unfortunately, the genuine benefits of ICT have not been provided widely in the field of agriculture. Hence, there is still a huge digital divide between farming and other industries. The natural resources such as water, energy, and fertilizers consumed by agricultural communities add more dimensions to the already existing problems such as global warming, soil degradation, and depletion of ground water. The agricultural community needs to be brought into the mainstream of technological innovations that could enable us to arrest the ever increasing global warming effect.

There is a strong requirement for the unified framework to provide seamless integration facilities to connect all the heterogeneous resources and adaptive information processing capabilities and also to convey required information to the farming community based on required context over a 360-degree view. This chapter discusses some generic frameworks which help to design and implement a context aware and adaptive information network. The design of a ubiquitous context aware middleware architecture for precision agriculture to solve major issues such as wastage of water, improper application of fertilizer, choice of wrong crops and season, poor yield, and lack of marketing is presented in this chapter. The method helps farmers throughout the farming lifecycle to choose crops, cultivate effectively, and sell their products. In this chapter the mobile sensor applications with the architecture of neural networks have been presented. As an example for the context aware services, a mobile health care service has been explained by utilizing the context over the physical and environmental sensor nodes.

Bibliography

[1] A. Z. Abbasi, N. Islam, and Z. A. Shaikh, A review of wireless sensors and network applications in agriculture. *Computer Standards & Interfaces*, 36(2):263–270, 2014.

[2] A. Baggio. Wireless sensor networks in precision agriculture. In *ACM Workshop on Real-World Wireless Sensor Networks, Stockholm,*. Citeseer, 2005.

[3] T. Basu, V. R. Thool, R. C. Thool, and A. C. Birajdar. Computer based drip irrigation control system with remote data acquisition system. 2006.

[4] N. D. R. Cheng, Software engineering for self-adaptive systems: research roadmap. In *Software Engineering for Self-Adaptive Systems*, Volume 5525 of *Lecture Notes in Computer Science*, pages 128–145. Springer, 2009.

[5] J. Cheng and R. Greiner. Learning Bayesian belief network classifiers: Algorithms and system. In *Advances in Artificial Intelligence*, pages 141–151. Springer, 2001.

[6] F. Chiti, R. Fantacci, F. Archetti, E. Messina, and D. Toscani. An integrated communications framework for context aware continuous monitoring with body sensor networks. *Selected Areas in Communications, IEEE Journal*, 27(4):379–386, 2009.

[7] D. L. Davies and D. W. Bouldin. A cluster separation measure. *Pattern Analysis and Machine Intelligence, IEEE Transactions*, (2):224–227, 1979.

[8] B. Fritzke. Some competitive learning methods. *Artificial Intelligence Institute*, Dresden University of Technology, 1997.

[9] J. Hwang, C. Shin, and H. Yoe. Study on an agricultural environment monitoring server system using wireless sensor networks. *Sensors*, 10(12):11189–11211, 2010.

[10] Y. Kawahara, H. Kurasawa, and H. Morikawa. Recognizing user context using mobile handsets with acceleration sensors. In *Portable Information Devices. IEEE International Conference* , pages 1–5. IEEE, 2007.

[11] Y. Kim, R. G. Evans, and W. M. Iversen. Remote sensing and control of an irrigation system using a distributed wireless sensor network. *Instrumentation and Measurement, IEEE Transactions*, 57(7):1379–1387, 2008.

[12] T. Kohonen. *Self-Organizing Maps*. Physics and astronomy online library. Springer, 2001.

[13] A. Krause, E. Horvitz, A. Kansal, and F. Zhao. Toward community sensing. In *Proceedings of 7th International Conference on Information Processing in Sensor Networks*, pages 481–492. IEEE Computer Society, 2008.

[14] A. Krause, D. P. Siewiorek, A. Smailagic, and J. Farringdon. Unsupervised, dynamic identification of physiological and activity context in wearable computing. In *Null*, page 88. IEEE, 2003.

[15] H. B. Lim, Y. M. Teo, P. Mukherjee, V. T. Lam, W. F. Wong, and S. See. Sensor grid: integration of wireless sensor networks and the grid. In *Local Computer Networks, IEEE Conference*, pages 91–99. IEEE, 2005.

[16] N. Malik, U. Mahmud, and Y. Javed. Future challenges in context-aware computing. In *Proceedings of IADIS International Conference WWW/Internet*, pages 306–310, 2007.

[17] S. Misra, V. Tiwari, and M. S. Obaidat. Adaptive learning solution for congestion avoidance in wireless sensor networks. In *Computer Systems and Applications. IEEE/ACS International Conference*, pages 478–484. IEEE, 2009.

[18] S. Mittal, A. Aggarwal, and S.L. Maskara. Online cleaning of wireless sensor data resulting in improved context extraction. *International Journal of Computer Applications*, 60(15):24–32, 2012.

[19] M. Mizoguchi, T. Ito, and S. Mitsuishi. Ubiquitous monitoring of agricultural fields in Asia using sensor network. In *Proceedings of 19th World Congress of Soil Science*, pages 125–128. International Union of Soil Sciences, 2010.

[20] D. T. Pham and G. A. Ruz. Unsupervised training of Bayesian networks for data clustering. In *Proceedings of Royal Society of London*, volume 465, pages 2927–2948. Royal Society, 2009.

[21] A. Sashima, Y. Inoue, T. Ikeda, T. Yamashita, and K. Kurumatani. Consorts-s: A mobile sensing platform for context-aware services. In *Intelligent Sensors, Sensor Networks and Information Processing, International Conference*, pages 417–422. IEEE, 2008.

[22] B. Schilit, N. Adams, and R. Want. Context-aware computing applications. In *Mobile Computing Systems and Applications, First Workshop*, pages 85–90. IEEE, 1994.

[23] A. Ultsch. Data mining and knowledge discovery with emergent self-organizing feature maps for multivariate time series. *Kohonen Maps*, 46:33–46, 1999.

[24] T. Wark, P. Corke, P. Sikka, L. Klingbeil, Y. Guo, C. Crossman, P. Valencia, D. Swain, and G. Bishop-Hurley. Transforming agriculture through pervasive wireless sensor networks. *Pervasive Computing*, 6(2):50–57, 2007.

[25] W. Zhang, G. Kantor, and S. Singh. Demo abstract: integrated wireless sensor/actuator networks in an agricultural application. *ACM SenSys*, Volume 4, 2004.

Chapter 9

Smart Algorithms for Energy-Aware Wireless Body Area Networks in Health Care Delivery

Ernesto Ibarra

University of Barcelona

Angelos Antonopoulos

Telecommunications Technological Centre of Catalonia

Elli Kartsakli

Technical University of Catalonia

Christos Verikoukis

Telecommunications Technological Centre of Catalonia

9.1 Introduction

In the past decade, the diagnosis, treatment and monitoring of patient health have been greatly improved by advances in biomedical technology. The design and implementation of medical devices such as pacemakers, insulin pumps and bionic prosthetics, among others, have expanded the ways to safeguard patient health and enhance quality of life. As technology evolves, these medical devices are becoming more efficient by getting smaller, more robust, and more comfortable for the patient. Furthermore, the use of wireless technologies in this field has led to the new paradigm of pervasive healthcare, enabling the remote monitoring and management of diseases, thus providing significant benefits for both patients and health providers.

This technological revolution along with the rising demand for health services constitute the preamble to the new concept in telemedicine, known as "mobile health" (mHealth). Wireless body area networks (WBANs) are the key enabling technology for the development of mHealth systems. WBANs consist of small and smart medical devices located in the vicinity of (on-body), or inside (in-body), the human body. The main purpose of WBANs is to efficiently manage the data communication among all the medical devices that belong to the network. Recently, a new standardization effort has developed the IEEE 802.15.6 standard for WBANs, for short-range ultra-low power wireless communications in the human body. Figure 9.1 shows the relationship between IEEE 802.15.6 and the other IEEE standards for wireless networks.

The WBANs will be the main driver in the growth of mHealth, increasing their functionalities and health benefits for the patients. The potential social and economic benefits of the implementation of WBANs in the current health systems are extremely promising. WBANs will allow patient diagnosis, treatment and monitoring, independently of their location, without interrupting their daily activities. This will improve both the effectiveness of medical procedures and the quality of life of patients.

The wireless interface consumes a considerable part of the energy available during the operation of the node, and the lack of energy may restrict the wireless connectivity of the nodes. Hence, the medium access control (MAC)

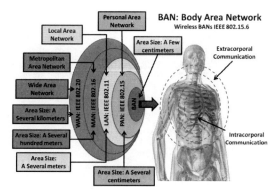

FIGURE 9.1: Relation between IEEE 802.15.6 and other wireless standards

sublayer is the most appropriate level to address the energy efficiency issues. The MAC layer defines the set of mechanisms and protocols through which various nodes agree to share a common transmission medium. In particular, the MAC protocols specify the tasks and procedures required for the data transfer between two or more nodes in the network, providing mechanisms to resolve potential conflicts that may arise during their attempt to access the medium.

Due to the space limitations of the human body, the task performed by each node is unique. For this reason, body nodes should be able to carry out their functions efficiently and interact with the human body in a discrete and undetectable way. To that end, body nodes should be small and light to adapt seamlessly to the human body. These characteristics strongly depend on the battery capacity, which is proportional to the battery's weight and size.

Apart from the physical limitations mentioned above, the limited battery capacity also restricts the lifetime of the nodes, since it is a finite power source. Hence, harvesting energy from other available sources could permanently supply the node with the required power, providing the most promising solution to the energy problem. Using special hardware devices, known as energy harvesters, body nodes can convert several types of energy present in the human environment (e.g., heat, motion, etc.) into electrical energy. A node powered by energy harvesting is able to collect energy in the human environment and use it for its own operation, thus achieving higher autonomy, overcoming the problems associated with the depletion of the traditional batteries.

This chapter is focused on the design of energy efficient solutions that exploit the energy harvesting capabilities of the wireless nodes. Our key objective is to describe the complete state of the art of WBANs, revealing various challenges for energy efficient management and quality of service (QoS) improvement. Therefore, we present efforts of the scientific community to identify and design energy efficient solutions capable of satisfying the QoS requirements of the WBANs.

We focus on three main trends in this field of research, which include MAC protocol design, smart energy-aware algorithms and energy harvesting-aided power schemes for WBANs. In addition, we present two novel solutions (i.e., an energy harvesting-aware MAC protocol and a smart power control algorithm) that further improve the energy efficiency and the performance of WBANs. Finally, we discuss the potential use of soft computing techniques to further enhance the proposed schemes, leading to low-complexity, efficient and practical solutions for mHealth applications.

9.2 Timeliness

Two unprecedented events in the history of humanity are now taking place: (i) the elderly population is greatly increasing worldwide, and (ii) mobile and wireless technologies are undergoing tremendous proliferation. With regard to the trend of global population aging, the World Health Organization (WHO) has recently foreseen that the number of people over 65 years old will soon be greater than the number of children under 5 years old [1]. According to WHO, the elderly population worldwide is growing by 2% every year due to the decreasing fertility rates and increased life expectancy. Furthermore, the growth rate of the elderly population is expected to reach 2.8% in 2030. Based on these facts, WHO warns that the rapid growth of the elderly population in some countries will challenge their national infrastructures and, particularly, their health systems.

Today, as a result of various factors (e.g., aging, lifestyle, diet changes, etc.), developing countries are experiencing an increase in the incidence of chronic diseases that mainly affect adults and elderly people [2]. Chronic diseases are characterized by long duration and generally slow recovery. The four main types of chronic diseases are: cardiovascular diseases, chronic respiratory diseases, diabetes, and cancer. These kind of diseases are mainly found in developed countries and constitute a major problem for the health system, creating the need to invest more resources in medication, hospital infrastructure, health workers, and medical devices. Often, these investments are not sufficient compared to the huge demand for services required by the population. An analysis made by WHO indicated that 23 countries with low and middle income levels will suffer economic losses because of chronic diseases and these illnesses will cost them around 83 billion dollars (approximately 63 billion euros) between 2006 and 2015 [1].

On the other hand, according to International Telecommunication Union (ITU) statistics, in 2014, the number of mobile subscriptions has almost reached that of the worldwide population (i.e., 7 billion mobile-cellular subscriptions worldwide). Penetration rates of mobile-cellular subscriptions are estimated to be around 96% globally, with 121% in developed countries and

90% in developing countries [3]. Moreover, ITU statistics have shown that more than 3 billion people are using the Internet this year, which corresponds to 40% of the population around the world. The high penetration rates of mobile-cellular networks in countries of low and middle income exceeds other civil and technological projects, such as healthcare infrastructure, paved roads and drinking water systems [4, 5].

The great advances in mobile and wireless technology in both developed and developing countries are surprising. This is the main reason that the scientific community and industry bet significantly on telemedicine systems based in wireless technologies.

9.3 Mobile Health (mHealth)

The budgets of health systems worldwide are greatly affected by the aging population and the increase in chronic diseases, since constant health monitoring and assistance to perform basic daily activities are often needed in these cases. In this context, the mHealth paradigm, which is part of electronic health (eHealth), is becoming increasingly popular and is potentially able to change the trajectory of health delivery. The Global Observatory for eHealth (GOe) defines mHealth as the medical and public health service carried out through mobile devices, such as cellphones, personal digital assistants (PDAs), patient monitoring devices, etc. [4].

mHealth has utilized the aforementioned technology platforms to perform various tasks related to the access, delivery, and management of patient information. Furthermore, it can provide the means to extend existing health care coverage to places with limited accessibility and significantly improve the quality of health care delivery in a wide variety of scenarios. Some indicative needs are maternal and child health care, general health services and information, clinical diagnosis, and treatment adherence, among others [4, 5].

WBANs will be the driving technology in the growth of mHealth, enhancing its functionality and increasing health benefits for the patients [6]. A typical mHealth system that employs WBAN technology is depicted in Figure 9.2. The potential social and economic benefits of the implementation of WBANs in the current health systems are extremely promising. WBANs will allow patient diagnosis, treatment, and monitoring independently of the patient's location without interrupting their daily activities. This will improve both the effectiveness of medical procedures and the patients' quality of life, while reducing the overall cost of health care delivery.

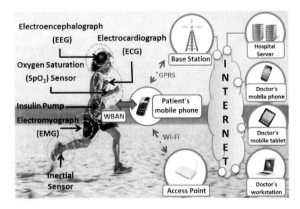

FIGURE 9.2: Example of WBAN in an mHealth system: the collected data are transmitted wirelessly to an external device (e.g., a smartphone or a PDA)

9.4   WBANs: Challenges and Open Issues

The main purpose of WBANs is to efficiently manage the data communication among all the medical devices that belong to the network. Nowadays, an increasing number of biomedical devices employ wireless connectivity for the exchange of information, to provide enhanced freedom for both patients and health workers. Hence, WBANs are a suitable candidate technology for the coordination of these wireless medical devices, called body nodes (BNs). Generally, BNs in a WBAN are heterogeneous, since they perform different tasks and have diverse QoS requirements and power requirements.

Several challenges should be overcome before the final implementation of WBANs in telemedicine systems. These networks usually face space constraints in order to adapt seamlessly to the human body, limiting the number and the size of the nodes in the network. The BN dimensions are strongly related to the batterys weight and size, which is proportional to the battery capacity. Since the battery is a finite source of energy, as the battery level drops, the BN operation becomes compromised and eventually stops. To resume operation, it is necessary to replace or recharge the battery as soon as possible. However, battery replacement is not always feasible, since it might damage the BN and even jeopardize the patient's health. This problem is exacerbated in the case of implantable nodes, where BN replacement would require surgical procedures [7, 8]. Hence, in an effort to prolong the lifetime of the network, there is an imperative need to exploit other sources of energy through energy harvesting techniques, in order to overcome the finite battery capacity, and to ensure the energy efficient operation of WBANs.

9.5 Body Sensor Nodes (BNs)

WBANs consist of small, smart medical devices with sensing, processing, and wireless communication capabilities. These BNs have the ability to act on their own without assistance from other devices. The nodes that belong to the WBANs may execute one or more actions related to physiological signal monitoring, diagnosis, or treatment of diseases. As seen in Figure 9.3, the BNs are heterogeneous with regard to their tasks (applications), hardware type, location in the human body, and propagation mediums of the radio wave (e.g., biological tissues, air, etc.). Figure 9.3 illustrates the heterogeneity of the BNs in the WBANs and shows some examples of BNs located:

On the human body: (a) electroencephalograph (EEG) (e.g., EEG IMEC or Holst centre), (b) respiratory rate sensor (e.g., Masimo rainbow SET acoustic monitoring), (c) electrocardiograph (ECG) (e.g., ECG V-patch), and (d) pulse oximeter (e.g., OEM fingertip pulse oximeter).

Inside the human body: (e) deep brain stimulator (e.g., LibraXP deep brain stimulator), (f) vagus nerve stimulator (e.g., Cyberonics VNS Inc), (g) subcutaneous implantable cardioverter-defibrillator (e.g., Cameron Health S-ICD system), and (h) gastric stimulation system (e.g., DIAMOND system).

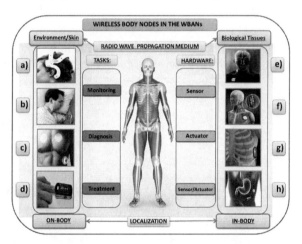

FIGURE 9.3: Body sensor networks

9.5.1 BN Characteristics

Due to the great heterogeneity of BNs and their respective characteristics, the nodes can be easily distinguished according to the following characteristics:

(i) **Priority**: The level of importance of each node. Since the clinical environment is extremely random, node priority will also be volatile since it depends on the current health status of the patient and the parameters studied over a given time.

(ii) **Quality of Service (QoS)**: This concept is related to the requirements in the treatment of specific data traffic. The parameters handled in QoS (e.g., packet loss, delay, and throughput) depend on the requirements of the executed application.

(iii) **Data Packet Size**: The sizes of data packets usually depend on the sensor type and the application.

(iv) **Inter-Arrival Packet Time**: The time delay between two consecutive packets from the same sensor.

(v) **Power Consumption**: The energy consumption of each BN per unit of time.

(vi) **Duty Cycle**: The percentage of usage of each component of the node (e.g., radio, CPU, or sensor) in a certain period. In the case of BNs, the duty cycle can be defined as the ratio of the active versus inactive periods [9].

(vii) **Battery Lifetime**: It represents the autonomy of the BN, which is the total operation time of a BN powered by batteries.

9.5.2   BN Energy Consumption

Energy consumption is key in wireless body area network (WBAN) design because of limited power supplies, operation of nodes, and their delicate functions which may be affected by power fluctuations or cessations. Figure 9.4 shows how medical WBANs are designed to handle data rates ranging from a few kilobytes per second up to several megabytes per second while consuming small amounts of energy (a few millijoules per second).

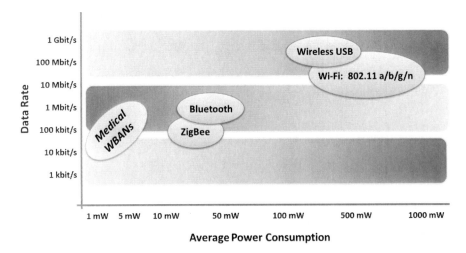

FIGURE 9.4: Medical WBANs: data transmission rates and average power consumption compared to other wireless technologies

As shown in Figure 9.5, the greatest amount of energy of an ECG node is consumed by the radio interface. The example shows that the radio uses 50% of the total energy consumed per unit of time, followed by the power management (PM) system at 25%, the sensor and reader (R-out) at 19%, the microcontroller unit (MCU) at 5%, and, finally, the analog-to-digital converter (ADC) at 1%.

Efficient energy management is the key to extending node lifetime as much as possible before it becomes necessary to replace or recharge their batteries. Most wireless nodes have several modes of operation supported by their MCUs. The modes are characterized by various power consumption levels due to internal switches that can turn nodes on or off, depending on the task performed. As a result, the energy consumption per unit of time can increase or decrease.

Figure 9.6 shows the power consumption of four operation modes of a wireless sensor, namely, active, sleep, reception, and transmission. The transmission and reception modes consume more power. However, the time a node spends in a specific mode and the number of transitions between modes greatly affects its duty cycle and energy consumption. The duty cycle of each node depends on the type of application and the tasks to be performed, which is why operating modes conform to these requirements.

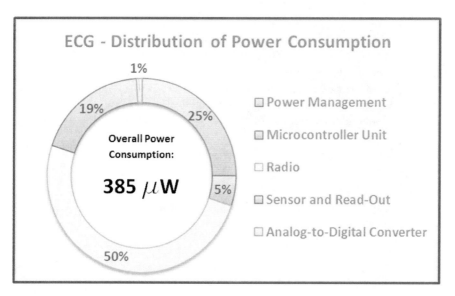

FIGURE 9.5: Distribution of power consumption when a sampled signal is transmitted by ECG [10]

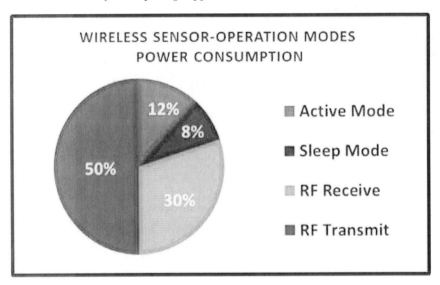

FIGURE 9.6: Power consumption of key operation modes in a wireless BN [11]

9.5.3   Energy Efficiency in WBANs

Wireless communication consumes a considerable portion of the energy available in the nodes. Therefore, the MAC layer is probably the most appropriate level to address energy efficiency issues [12]; it carries out functions related to channel access control, scheduling of the transmissions, data framing, error handling, and energy management, among others. An efficient MAC protocol maximizes the data throughput and the energy efficiency of the network, thereby achieving optimum use of the wireless channel and extending the lifetime of the batteries.

In wireless networks, sharing the medium involves situations that increase the power consumption and therefore reduce the lifetime of the network. Such energy consuming events are [13]:

Packet Collision: This occurs when multiple packets are transmitted at the same time. The retransmissions of collided packets require additional energy consumption.

Idle Listening: This happens when a node spends time listening to an idle channel when waiting to receive data.

Overhearing: This occurs when a node listens to the channel to receive packets destined for other nodes.

Packet Overhead: This refers to the control packets and the information added to the payload (headers). The number of control packets used to carry out data communication also increases power consumption.

9.5.4 Smart Energy-Aware Algorithms for BNs

WBANs have more stringent QoS requirements with respect to traditional wireless sensor networks (WSNs) [14, 15]. Given the strict demands for low delay and packet loss in WBANs, Kateretse et al. [16] proposed a traffic differentiation and scheduling scheme based on patients' data classification and prioritization according to their current status and diseases.

Several authors also focused on enhanced routing schemes, in order to improve QoS performance. Liang and Balasingham [18] proposed a novel QoS-aware framework for prioritized routing user specific QoS support in WBANs. The QoS requirements and the specific demands of the BNs were determined based on the patient's health conditions, the clinical application, and the characteristics of the collected data. Hassanpour et al. [19] introduced a routing scheme based on genetic algorithms to improve reliability and delay. Finally, Tsouri et al. [20] proposed a global routing scheme with a novel link cost function designed to balance energy consumption across the WBAN to increase network lifetime.

More complex approaches have also been proposed. For instance, Razzaque et al. [17] presented a data-centric multi-objective QoS-aware routing (DMQoS) protocol for WBANs, based on a modular architecture consisting of five modules: the dynamic packet classifier, delay control, reliability control, energy-aware geographic routing, and QoS-aware queuing and scheduling modules. DMQoS addresses QoS issues for multihop WBANs, focusing on enhanced delay and reliability. Otal et al. [21] designed and implemented a novel QoS fuzzy-rule-based cross-layer scheduling algorithm for WBANs, aiming to guarantee a specific bit-error-rate (BER) for all packets, within particular latency limits and without endangering the battery life.

9.5.5 BNs Powered by Human Energy Harvesting (HEH)

In recent years, the interest of the scientific community and electronic industry in energy harvesting as a power source in low-power wireless devices has significantly increased [22], since it is projected to be an ideal solution for eliminating the energy dependence of electronic devices on batteries. The harvesting process is carried out through a device called an energy harvester. This device transforms a physical or a chemical source present in the environment into electrical energy.

Energy harvesting sources are not uniformly distributed in the human environment, i.e., the magnitude and availability of a source depends on its characteristics. In this way, certain phenomena are more prevalent in some parts of the body. A complex tradeoff must be considered when designing energy harvesters for wireless nodes [22], taking into account the characteristics of the sources to be harvested, the energy storage devices, the power management of the nodes, the employed communication protocols and the application requirements.

Due to the heterogeneous components of WBANs, the selection of the best type of energy harvester for each BN is an important task. Generally, BNs differ in power consumption requirements, depending on their target medical application, the monitored physiological signal, the duty cycle, etc. Hence, the performance of the energy harvester must match the energy demands of the BN, in order to ensure continuous operation and extended lifetime. Furthermore, the positioning of BNs on or inside the human body poses additional limitations, since different energy sources may be available for harvesting in each case.

9.6   Medium Access Control (MAC) Mechanisms

MAC protocols are divided into contention-free and contention-based access schemes. In contention-free protocols, each node is scheduled to access the medium without interfering with other nodes. This can be done either by direct assignment from the central controller through a polling request (polling-based scheme) or by assigning different time slots (TDMA or time division multiple access), different frequency channels (FDMA or frequency division multiple access), or different unique codes (CDMA or code division multiple access). Figure 9.7 illustrates TDMA, FDMA, and CDMA contention-free methods. Due to energy efficiency requirements, medium access schemes based on FDMA and CDMA are not feasible solutions for WBANs, since they require complex hardware circuitry and high computational power [23, 24].

Contention-based protocols provide random access to the medium and the nodes must compete for channel resources. These protocols fall into two sub-categories, depending on whether they include mechanisms for checking channel availability before transmitting data (e.g., CSMA/CA or carrier sense multiple access/collision avoidance) or lack such mechanisms (e.g., ALOHA).

The nodes in CSMA/CA perform clear channel assessment (CCA) by listening to the wireless channel and transmitting only if the medium is perceived as idle. If the channel is busy, the node postpones its transmission until the channel becomes idle. The receiver sends an acknowledgment (ACK) to the transmitting node if the transmission is successful. In ALOHA, nodes attempt transmissions as soon as their buffers have packets. If a data collision occurs, each node retransmits after a random time interval to reduce the probability of further collisions.

In slotted protocols, time is divided into discrete intervals (slots) and actions (transmitting or receiving packets) are initiated at the beginning of each slot. This reduces packet collisions significantly and improves network performance. Nodes plan and use the time slots depending on the duty cycle of the radio, i.e., whether it is in reception mode, transmission mode, or turned off. The slots must be synchronized among all the nodes, allowing them to turn

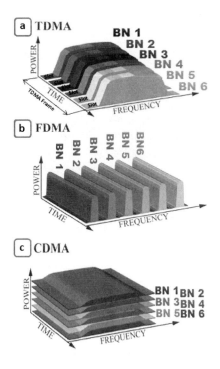

FIGURE 9.7: Contention-free multiple-access methods: (a) TDMA, (b) FDMA, and (c) CDMA [25]

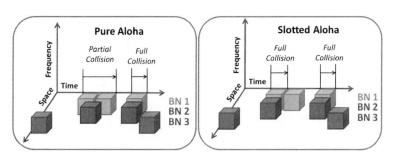

FIGURE 9.8: Timing diagram of pure and slotted ALOHA [26]

their radios on only when needed, thereby greatly reducing the idle listening. An example of such a protocol is slotted ALOHA, which is more suitable for WBANs due to its enhanced medium utilization efficiency with respect to pure ALOHA (see Figure 9.8).

In contention-based MAC schemes, the main advantages are good scalability, adaptation to traffic load fluctuations, low complexity, lower delay, and reliable transmission of packets. However, their transmission efficiency is reduced due to packet retransmissions, while the power consumption is relatively high due to overhearing, idle listening, and, above all, packet collisions [23]. Furthermore, as shown experimentally in [28], CSMA/CA-based protocols also have problems with unreliable CCA, especially in the case of in-body WBANs where there is rapid attenuation of the electromagnetic waves through biological tissues. With respect to pure and slotted ALOHA, Javaid et al. [26] explain that these techniques are not widely used in WBANs due to their high packet drop rates and low energy efficiency due to data collisions.

On the other hand, the TDMA-based protocols have high transmission efficiency, no overhearing problems, no packet collision, low power consumption, and maximum bandwidth utilization. In analytical studies [26, 29], the authors considered the TDMA-based protocols as the most energy efficient and reliable MAC protocols for WBANs. Their main disadvantage is the lack of flexibility and scalability, while they require precise synchronization (unlike CSMA/CA and ALOHA).

In terms of mobility, the CSMA/CA approach can be better adapted in dynamic environments, with respect to TDMA, thereby providing good mobility support to the wireless network [24]. On the contrary, TDMA schemes are more appropriate for static WBANs, as they can support high traffic volumes.

9.6.1   MAC Protocols for WBANs

The design of energy efficient MAC protocols for WBANs has captured the interest of both the research community and the industry, in recent years. This has led to the development of numerous MAC protocols for WBANs, aiming to improve the energy efficiency of the network and to extend the battery lifetime of the BNs.

With the aid of the scientific and industrial community, the IEEE Task Group 6 published on February 29, 2012 the IEEE 802.15.6 standard for WBANs [30]. The protocol provides specifications and recommendations for the physical (PHY) and MAC layers for WBANs. In IEEE 802.15.6, time is divided into superframes. This structure allows three types of access mechanisms: (i) random access (contention-based), which uses CSMA/CA or slotted ALOHA for resource allocation, (ii) improvised and unscheduled access, which uses unscheduled polling and posting for resource allocation, and (iii) scheduled access, which schedules the allocation of slots in one (1-periodic) or multiple (m-periodic) time allocations.

In addition to the standard, several surveys of proposed MAC protocols can be found in the literature [23, 24, 26, 29, 31]. In particular, a comprehensive review of the key PHY and MAC design approaches and challenges is given in [31], especially focusing on end-to-end architectures. In the remaining of this section, a review of existing MAC protocols for WBANs will be given, discussing their key features.

TDMA-based schemes have been widely adopted for WBAN applications, due to their high energy efficiency, which is a crucial requirement in WBANs in order to extend the lifetime of the BNs. Timmons and Scanlon [32] introduced a TDMA-based protocol called medical MAC (MedMac), which does not require any synchronization overhead. The synchronization of the nodes is maintained through a combination of timestamp scavenging and an innovative adaptive guard band algorithm. In [33], Fang and Dutkiewicz proposed another TDMA-based MAC protocol (BodyMAC) that uses flexible and efficient bandwidth allocation schemes and sleep mode to work in dynamic applications in WBANs. Li and Tan [34] proposed a heartbeat-driven MAC (H-MAC) protocol. H-MAC is also based on TDMA, but it uses the rhythm of the heartbeat to perform time synchronization, eliminating the energy expenditure of synchronization to prolong network life significantly.

In the extremely volatile clinical environments, the importance of each BN depends on the current health condition of the patient and the parameters studied at any given time. For this reason, adding and removing nodes in a fast and easy way is a desirable feature in WBANs. This flexibility is supported by polling schemes, which, in general, have two main advantages: (i) deterministic and bounded transmission delay, and (ii) scalable network architecture [35]. In a typical polling network, when a node is willing to enter the network, it sends a JOIN message to the coordinator. Upon receiving this message, the coordinator verifies the request and creates new polling and data slots for the new node. If a BN wants to leave the network, a DEPART request is send, and the coordinator proceeds to remove the allocated resources.

Khan and Yuce [35] explain that the polling-based MAC protocols can support traffic sources with different data inter-arrival rates, providing higher network flexibility than TDMA-based protocols. They also claim that the combination of polling-based and CSMA/CA-based access protocols could be a good mechanism for power saving and reliable communication of critical medical data.

Boulis and Tselishchev [36] studied the performance of contention-based and polling-based access under different traffic loads in WBANs. Their results indicate that significant energy gains can be obtained through polling. Regarding the latency (end-to-end delay), the combination of short contention periods with long polling periods provides the most stable performance for packet transmissions.

The use of secondary wake-up radios has also been considered as a means reduce energy consumption in BNs. Ameen et al. [37] designed a MAC protocol using TDMA combined with an out-of-band (on-demand) wakeup radio through a centralized and coordinated external wakeup mechanism. The communication process takes place in two stages: (i) a wakeup radio is used to activate the node and (ii) a main radio is used for control and data packet exchange. The coordinator maintains a table with the wakeup scheduling of every node in the network, constructed according to the network traffic, while the wakeup intervals are calculated by the packet inter-arrival time. The authors proved through extensive simulations that their method outperforms well-known low-power MAC protocols for WSNs, including B-MAC [38], X-

MAC [39], WiseMAC [40] and ZigBee (power saving mode) [41] in terms of energy efficiency and delay. This MAC protocol depends on a fixed, predetermined schedule, which has two side effects: (i) the state changes of the nodes from active to inactive mode may cause idle slots in the system, and (ii) flexibility is restricted since it is cumbersome to add or remove nodes in the network.

Since WBANs can support a wide range of applications with different requirements, it is important to design MAC protocols that can provide QoS guarantees. Otal et al. [42] proposed the distributed queuing body area network (DQBAN) MAC protocol. DQBAN divides TDMA slots into smaller units of time (called minislots) for serving access requests, while the data packets use the normal slots. To satisfy the stringent QoS demands in WBANs, DQBAN introduces a novel cross-layer fuzzy-logic scheduling mechanism and employs energy-aware radio activation policies to achieve a reliable system performance. The on-demand MAC (OD-MAC) protocol is another interesting approach proposed by Yun et al. [43]. OD-MAC is based on IEEE 802.15.4 with some modifications to support WBAN requirements such as real-time transmission, collision avoidance, and energy efficiency. It uses guaranteed time slots to satisfy real-time transmissions and collision avoidance while it adjusts the duration of the superframes to provide better energy efficiency.

More recent works have also considered the possibility of cloud-based MAC layer coordination, in order to optimize the use of resources and enhance QoS, reliability and energy consumption [44,45]. Such solutions are particularly suitable for heterogeneous deployments of body and ambient sensors in complex multihop topologies that can often be found in ambient assisted living environments. The proposed schemes support simultaneous transmissions from multiple BNs with network coding capabilities and use centralized scheduling in order to increase the probability of correct decoding at the receiver.

9.6.2   MAC Protocols for HEH-Powered WBANs

Currently, the majority of MAC protocols proposed for WBANs aim to improve QoS and optimize power consumption through efficient scheduling and battery management. However, the difficulties introduced by power sources based on energy harvesting are not effectively addressed. In the literature, papers model and analyze energy harvesting in WBANs using probabilistic models based on Markov chains [46,47], optimal numerical solutions for energy efficient transmission strategies [48], and resource allocation [49,50]. Nonetheless, none of the aforementioned works proposes a MAC protocol that supports energy harvesting techniques.

The main challenge in energy harvesting-based WBANs is to design protocols that provide access depending on the BN priorities, taking into consideration their particular energy supply conditions. Eu et al. proposed a MAC protocol specially designed for WSN powered by ambient energy harvesting (WSN-HEAP), known as probabilistic polling, for both single-hop [51] and multi-hop [52] networks. The results show that the protocol is able to adapt to different factors such as changes in the energy harvesting rates and the number of nodes. Furthermore, due to this dynamic adaptation, throughput, fairness, and scalability can be substantially improved.

As we mentioned above, probabilistic polling has the ability to adapt network operation to fluctuations in both energy supply and node number. However, the application of such schemes to WBANs operating with energy harvesting is not straightforward, since the different energy levels of the BNs must be considered. Furthermore, probabilistic access does not ensure prioritization of BNs, which is crucial to ensure the early detection of important events concerning the patient's health. In the following section, we introduce our proposed solution to tackle these issues while exploiting the energy harvesting capabilities of the BNs.

9.7   Novel MAC Protocol Design for HEH-WBANs

In WBANs, throughput maximization, delay minimization, and lifetime extension of the network operation are some of the main goals to be achieved [53]. Even though energy harvesting can extend the lifetime of a WBAN, it may degrade other QoS metrics such as throughput, delay, and packet loss [54].

The measurement accuracy of the sensed data is another major challenge in WBANs. Unlike WSNs, where a large number of nodes can compensate for the lack of measurement precision, each BN in WBANs has a unique function, and should be robust and accurate. At the present time, WBAN-oriented MAC protocols take into account battery-powered BNs and are not compatible with energy harvesting-oriented networks. Hence, it is critical to propose and design MAC mechanisms for WBANs powered by human energy harvesting (HEH-WBANs), in order to guarantee an acceptable QoS level and make optimal use of the limited energy collected in the human environment.

The MAC design for HEH-WBANs should take into account the particular types and features of the power sources to be harvested in the human environment. It should also provide the nodes with medium access according to their priority and available energy. Taking into account recent developments in WBANs and in the energy harvesting field, this chapter section shows our proposal for efficient energy harvesting-aware resource management techniques that aim at a better QoS in HEH-WBANs.

Our first contribution is a hybrid MAC protocol called HEH-BMAC, which has been designed and developed for HEH-WBANs. HEH-BMAC uses a dynamic scheduling algorithm to combine user identification (ID) polling and probabilistic contention (PC) random access, adapting the network operation to the random, time-varying nature of the human energy harvesting sources. Moreover, it offers different levels of node priorities (high and normal), energy-awareness, and network flexibility. To the best of our knowledge, this is the first contribution to MAC protocol design for WBANs powered by energy harvesting in the human environment.

The second contribution of this research is a power-QoS control scheme, called PEH-QoS, for BNs powered by human energy harvesting. This scheme was designed to achieve the optimal use of collected energy and a substantial improvement in QoS. It intends to ensure that a node can both capture and detect medical events and transmit the respective data packets efficiently. One of the main features of our mechanism is that only useful data sequences are transmitted, discarding data packets that lost their clinical validity (i.e., out of date data).

9.7.1 A Hybrid Polling HEH-Powered MAC Protocol (HEH-BMAC)

In this section, we will present HEH-BMAC protocol for WBANs. First, we will describe the system model, followed by a complete description of the protocol rules and an operational example. We will show the efficiency of the proposed scheme through a thorough performance evaluation and outline the key conclusions of our work.

9.7.1.1 System Model

We adopt a star topology, where the head (sink) is the body node coordinator (BNC) responsible for setting up the network and collecting all the information transmitted by a number of lightweight and portable BNs. The BNs have different functionalities and different traffic loads (i.e., packet inter-arrival time and packet payload). Figure 9.9 illustrates the topology.

The events detected by the BNs can be signals carrying sensitive and vital information (e.g., ECG, EEG, etc.) or signals with random characteristics (e.g., motion, position, temperature, etc.). To model a realistic scenario based on the above arguments, we adopt the same inter-arrival times (IATBN) as in [55] for event generation.

In HEH-BMAC, each sensor is connected to an energy harvester. We assume that the BNC has an external power supply and higher processing capabilities than BNs, while the BNs have a constant energy harvesting rate (KEH). The energy harvester must be able to harness the available energy at all times and during all states of the node (i.e., sleep, idle, transmission, reception, and inactive states). The performance of the energy harvester directly affects the operation of the node, but not vice versa.

9.7.1.2 Protocol Overview

To our knowledge, HEH-BMAC is the first MAC protocol designed to adapt to different energy conditions introduced by human energy harvesting sources in WBANs. HEH-BMAC has two operation modes: (i) contention-free ID polling,

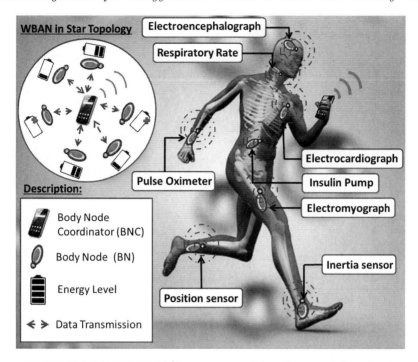

FIGURE 9.9: HEH-WBAN: system model and network topology

and (ii) probabilistic contention (PC) channel access. Hence, our protocol offers two levels of priority depending on the BN type.

The use of contention-free ID polling access is provisioned for nodes with predictable energy sources or nodes with high priority (ID-BNs). On the other hand, the use of contention-based PC access applies to nodes with unpredictable energy sources or nodes with normal priority (PC-BNs). The base of our protocol is an algorithm that performs time allocation in a dynamic way. The goal of the dynamic scheduling is to assign time periods for both ID polling and PC access. Due to the combination of these two access modes and the dynamic scheduling algorithm, the HEH-BMAC protocol is able to adapt to changes in network size and energy harvesting rate (K_{EH}). In the following subsections we describe the operation and different modes of our protocol:

- The proposed protocol offers service differentiation by combining two access methods: reserved polling access (ID polling) for nodes of high priority and probabilistic random access (PC) for nodes of normal priority.

- The ID and PC periods are dynamically adjusted according to the energy levels of the wireless nodes. HEH-BMAC facilitates flexibility by allowing the dynamic addition and removal of wireless sensor nodes.

ID Polling Access Mode:

All nodes in the HEH-WBAN are assigned a unique ID for data security, data control, and medical application. Figure 9.10a illustrates the communication process in ID polling mode which consists of three steps: (i) the BNC transmits a polling packet containing the ID of the BN to be polled, (ii) the polled BN responds with a data packet transmission, and (iii) the BNC sends an ACK packet that confirms the successful reception of the data packet. As shown in Figure 9.10b, the ID-BN remains in the sleep state until its turn to transmit. Upon reaching its turn, it wakes up and goes into the R_x state to receive the ID polling from the BNC. Once the communication is completed, the ID-BN turns its radio off until the next round of polling.

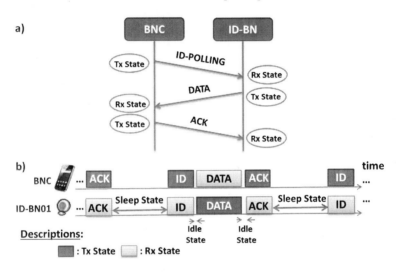

FIGURE 9.10: ID polling access mode: (a) data communication process and (b) ID-BN states and transmission

Probabilistic Contention (PC) Access Mode:

The PC access mode deals effectively with contention, achieving high through-put, and maintaining fairness for single-hop networks. In addition, this mode offers the advantage of adaptation to the changes in the energy harvesting rates, node failures, or additions and removals of nodes. In PC access [51,52], instead of ID polling, the BNC broadcasts a control packet (CP packet) that includes the value of the contention probability (CP). When a PC-BN (node in PC access mode) receives the CP packet, it generates a random number X_i, where $X_i \in [0, 1]$ and i is an integer identifier of the node.

If the value of X_i is less than that of the CP, the PC-BN transmits its data packet (Figure 9.11a); otherwise, the node transits to the idle state, waiting for the next CP packet (Figure 9.11b).

The CP is dynamically adjusted at the BNC according to an updating algorithm that takes into account the network load (traffic load and addition

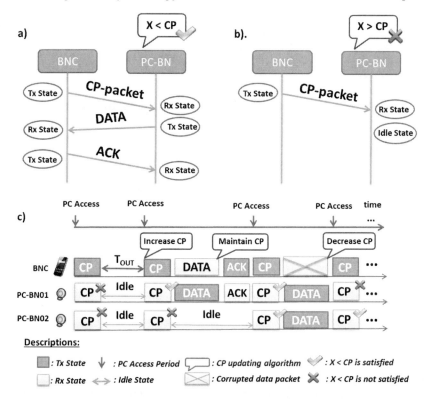

FIGURE 9.11: PC access mode: (a) data communication process when X<CP is satisfied, (b) data communication process when X<CP is not satisfied, and (c) CP updating algorithm and transmission process

or removals of nodes) and the K_{EH}. The value of the CP is updated if no PC-BN responds to the CP packet and the BNC increases the value of the CP threshold to increase the transmission probability of the PC-BNs. Second, when there is a collision between two or more PC-BNs, the BNC decreases the value of the threshold to reduce the probability of collision. In case of successful transmissions, the current value of the threshold is maintained in the next CP packet.

There are different techniques that can be employed to select the value of the contention probability [51]: (i) multiplicative increase-multiplicative decrease (MIMD), where the CP is modified exponentially, and (ii) additive increase-multiplicative decrease (AIMD), where the CP is increased linearly and decreased exponentially, and (iii) multiplicative increase-additive decrease (MIAD), where an exponential increase and a linear decrease of the CP is selected.

We use the AIMD technique because it provides higher throughput than the other schemes for single-hop scenarios. In AIMD, the CP is increased gradually by a factor $\alpha_{IN}(0 < \alpha_{IN} < 1)$ when polling is unsuccessful because of idle slots (i.e., $CP_{(t+1)} = CP_t + \alpha_{IN}$. In case of collisions, the CP is decreased by a larger factor $\beta_{MD}(0 < \beta_{MD} < 1)$(i.e., $CP_{(t+1)} = CP_t \times \beta_{MD}$).

An example is shown in Figure 9.10c. The BNC broadcasts a CP packet containing the contention probability that determines whether the PC-BN should transmit its data packet. In the case of no packet reception, the BNC waits for a predefined time-out period (T_{OUT}), updates the CP packet with the increased threshold and broadcasts the new value in the next PC round. The PC-BN only transmits its data packet if $X_i < CP$. If only one node transmits in the current PC round, the BNC sends the ACK packet to the polled PC-BN (successful transmission). In the case of packet loss (unsuccessful transmission) due to collision between two or more PC-BNs, the BNC updates the CP packet with the decreased threshold and the nodes are prepared to retransmit their data in the following PC round. All PC-BNs maintain a buffer to store the data to be retransmitted.

Dynamic Schedule Algorithm:
The BNC is responsible for allocating the ID polling periods and the PC access periods. The boundaries of these time periods are defined by the dynamic schedule algorithm, whose operation is illustrated in Figure 9.12. The algorithm performs two main tasks. The first is to assign a monitoring time ($T_{\alpha i(n)}$) and calculate the duration of the data communication process ($T_{\gamma i(n)}$) for each ID-BN i during the nth access period. The value of $T_{\alpha i(n)}$ for an ID-BN i must be greater or equal to the minimum time required to obtain enough energy to send its data packet. BNC calculates the $T_{\alpha i(n)}$ for each ID-BN using the K_{EH} of the respective harvester and the IAT_{BN} of the sampled sensor data. Thus, the BNC can estimate when each ID-BN has data to transmit and whether its energy level is sufficient to select an appropriate value for $T_{\alpha i(n)}$. On the other hand, $T_{\gamma i(n)}$ is defined as the time required for ID-BN i to perform a successful transmission. We also define $T_{\beta i(n)}$ as the moment of completion of the current communication process, i.e., $T_{\beta i(n)} = T_{\alpha i(n)} + T_{\gamma i(n)}$.

The current values of $T_{\alpha i(n)}$ and $T_{\beta i(n)}$ for all ID-BNs are stored in a dynamic table and employed by the BNC to coordinate the ID polling process. When the BNC proceeds to perform ID polling, the dynamic table is updated with the next values of $T_{\alpha(n+1)}$ and $T_{\beta(n+1)}$ for the next ID-BN to be polled. In this way, we can predict the responsiveness to an ID polling for a given node, and make the decision to poll it or not (thus improving the scalability of the system, since the dynamic table is constantly updated). Figure 9.12a presents an example of the dynamic calculation of ID polling and PC access periods operating together.

The second task of the dynamic schedule algorithm is to calculate the interval between two adjacent ID polling periods, for example (Figure 9.12b) the $T_{\beta i(n)}$ of the node $ID_{i(n)}$ and the $T_{\alpha j(n)}$ of the next node $ID_{j(n)}$. The BNC performs the calculation of this interval using the data provided in the dynamic table. If the time between two consecutive ID polling periods is sufficient for

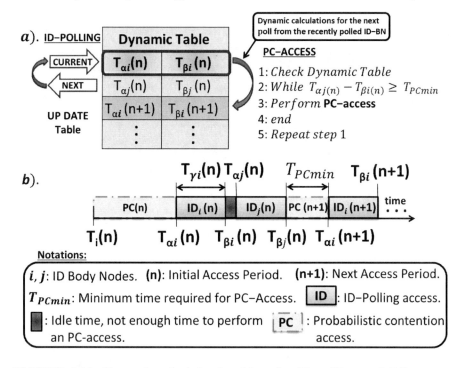

FIGURE 9.12: Dynamic schedule algorithm for ID polling and PC access periods in HEH-BMAC

a successful data transmission of a PC-BN in PC access mode $(t \geq T_{pC_{min}})$, this time is exploited for probabilistic contention (PC period). Otherwise, if this time is not sufficient $(t < T_{pC_{min}})$, the BNC remains idle, waiting for the next ID polling period.

9.7.1.3 Operation Example

Figure 9.13 shows an example of the HEH-BMAC protocol running on a network with four nodes, where two nodes are in ID polling mode and two are in PC access mode.

The protocol works as follows:

- The BNC performs the configuration and time calculations for the ID-BNs. The BNC stores the values $T_{\beta(ID-BN)}$ and the current values of $T_{\alpha(ID-BN)}$ for ID-BN01 and ID-BN02 in a dynamic table.

- At instant T_1, BNC initiates ID polling access for ID-BN01. Once the communication process has been completed, BNC updates the dynamic table with the next value of $T_{\alpha(ID-BN01)}$. ID-BN01 goes into sleep mode until its next ID polling period. ID-BN02 remains in sleep state waiting

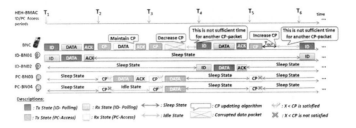

FIGURE 9.13: Frame exchange in HEH-BMAC

for its ID polling period. The PC-BNs remain in sleep state since they do not have packets for transmission.

- BNC uses the dynamic list to calculate the interval between two adjacent ID polling accesses (T_1 and T_4 in this example). In this example, the interval is sufficient for two successful data transmissions in PC access.

- At instant T_2, BNC sends the CP packet (starting PC access) to all PC-BNs (i.e., PC-BN03 and PC-BN04). In this example, PC-BN03 randomly selects $X_3 < CP$ whereas PC-BN04 selects $X_4 > CP$. Hence, PC-BN03 gains access to the medium and starts its data transmission, whereas PC-BN04 remains in idle state waiting for the next PC access period. The CP updating algorithm maintains the current threshold value.

- At instant T_3, BNC sends the next CP packet to all PC-BNs. The condition $X_i < CP$ is satisfied for both PC-BN03 and PC-BN04 and both nodes transmit their data packets, resulting in a collision. According to the CP update algorithm, the BNC must decrease the CP threshold and include the updated value in the next CP packet. In this example, the remaining interval (after the packet collision) is not sufficient for another PC access. Therefore, the BNC remains idle until the next ID polling period (which starts at T_4).

- At instant T_4, the BNC starts ID polling access for ID-BN02. Once the data transmission has been completed, the BNC updates the dynamic table with the next value of $T_{\alpha(ID-BN02)}$. Meanwhile, ID-BN01 is in sleep state waiting for its ID polling period. The table is used to calculate the next interval between T_4 and T_6 and determine whether there is enough time for PC access (in this example, the interval is sufficient for one successful data transmission in PC access).

- At instant T_5, BNC broadcasts the CP packet containing the new threshold value to all PC-BNs. In this example, neither PC-BN03 nor PC-BN04 selects a X_i that satisfies the condition $X_i < CP$ and neither node transmits in the current PC access.

The BNC waits for a predefined T_{OUT} and then increases the CP threshold value. Since the remaining interval is not sufficient for another PC access, the BNC remains idle until the next ID polling period (T_6).

9.7.1.4   HEH-BMAC with Energy Saving

HEH-BMAC is energy-aware, since it has been designed to operate in energy harvesting conditions. In particular, the behavior of each BN dynamically adapts to its energy level. The energy level of a node at a given moment can be defined as the energy stored in the battery plus the harvested energy minus the energy consumed by the radio interface. The modifications that energy-awareness brings to our protocol are explained next:

- **ID Polling Energy-Aware**:

  Dynamic schedule: The BNC calculates the $T_{\alpha(ID-BN)}$ based on the K_{EH} and IAT_{BN} of each ID-BN. The $T_{\alpha(ID-BN)}$ does not have a fixed value, since this time interval is continuously updated in the dynamic table to know in advance the energy state of a node at any given time. Thusly, we can predict the future responsiveness of a given node to an ID polling, and decide whether to poll it or not.

  Polling-awareness: When a node receives an ID poll packet, it checks its energy level. If the level is not sufficient, the node does not respond to the poll but enters a sleep mode. The BNC assigns the time reserved for this ID polling to the PC access users.

- **PC Access Energy-Aware**:

  Energy-awareness: The PC-BNs check their energy levels and data packet buffers to decide whether to participate in the PC access. If their energy is below a certain level or their buffers are empty, they enter sleep mode. All PC-BNs will be in sleep state during the ID polling.

  Polling-awareness: The PC access mode is employed if there is enough time between successive ID pollings. The BNC dynamically adjusts the CP packet according to the responses of the PC-BNs (through the CP updating algorithm).

9.7.1.5   Performance Evaluation

Simulation Consideration and Setup

In our simulation model, we assume a star topology for a network consisting of a BNC, K nodes in ID polling mode and L nodes in PC access mode. The simulation scenario is depicted in Figure 9.14. The nodes in our simulations are typical medical sensors, whose traffic characteristics and priorities are shown in Table 9.1. Let us recall that the high and normal priorities correspond to ID polling and PC access mode, respectively. The characteristics of the selected

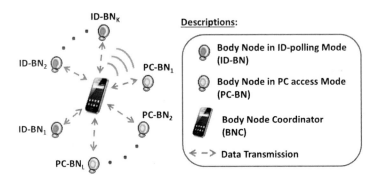

FIGURE 9.14: Simulation scenario

nodes in our experiments can be found in [55]. However, in the case of the ECG and blood pressure nodes, we adopt a slightly different aggregate traffic model which results in sample sizes of 120 bits and 96 bits, respectively. For this process, the bit rate and the delay requirements of health care data were taken into account [56, 57].

TABLE 9.1: BN Simulation Parameters

BN Type	IAT (ms)	Payload (bits)	Priority	Access
ECG	20	120	High	ID Polling
Respiratory rate	50	12	High	ID Polling
Blood pressure	80	96	High	ID Polling
Blood flow	25	12	High	ID Polling
Other signs	65	12	Normal	PC Access

The configuration parameters of the network were selected according to the IEEE 802.15.6 PHY-MAC specification for CSMA/CA operation [30], as shown in Table 9.2. The physical layer frame structures used (based on IEEE 802.15.6) are shown in Figure 9.15.

We assume that the ID-BNs perform data transmission in real time (no packet retransmissions and no packets stored in the buffer). In the PC-BNs, retransmissions may take place when a collision occurs (packets are stored in the buffer). However, when the energy level of a node is very low (almost zero), the node cannot proceed to the transmission or retransmission of packets, and packet loss may occur. Data packets that remain in the buffer after the simulation are considered lost packets. Moreover, the BNC maximum waiting time (T_{OUT}) for a response from the nodes (ID-BNs or PC-BNs) is assumed to be equal to the value of the short inter frame space period ($pSIFS$) 0.05

TABLE 9.2: MAC Protocol Parameters

HEH-BMAC		IEEE 802.15.6		
CP Updating Method		**User Priority**	CW_{min}	CW_{max}
Method	AIMD	High (UP_6)	2	8
δ_{add}	0.01	High (UP_3)	8	16
δ_{md}	0.5	High (UP_0)	16	64

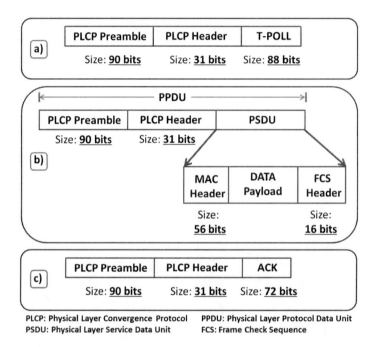

FIGURE 9.15: Physical layer frame structures: (a) polling packet, (b) data packet, and (c) ACK packet

ms. For the AIMD CP updating algorithm, we use $\alpha_{IN} = 0.01$ and $\beta_{MD} = 0.5$ since these values give high throughput for single-hop scenarios [51].

We assume that each node has incorporated an energy harvester that supplies power at a constant rate $K_{EH} = 1.3$ mJ/s. This value of K_{EH} allows our protocol to reach 100% throughput in the initial setup of the network. To evaluate our approach, we compared the performance of HEH-BMAC with the IEEE 802.15.6 standard protocol. For comparison, we used the data transmission rate of 485 kb/s, and adopted the non-beacon mode without the superframes of the IEEE standard. We chose this configuration because it offers

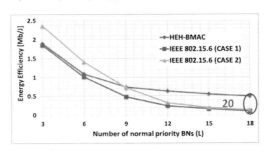

FIGURE 9.16: HEH-BMAC vs. IEEE 802.15.6: energy efficiency

random and prioritized access through the contention window (CW) bounds of CSMA/CA.

For the normal priority BNs we used the CW values that correspond to the user priority UP_0 of the IEEE 802.15.6. As for the high priority BNs we examined two different cases to evaluate IEEE 802.15.6 performance for different CW values and provide a fair comparison with HEH-BMAC. In Case 1, the CW values of the UP_6 were employed, whereas in Case 2, the CW values of the UP_3 were selected. Initially, the considered setup consisted of 4 BNs in high priority and 3 BNs in normal priority. The number of normal priority BNs was gradually increased up to 18 nodes. The system setup used to compare the performance of our protocol with the standard is shown in Table 9.2.

The metrics for the evaluation of the performance of our protocol are energy efficiency and normalized throughput. Energy efficiency was defined as the total amount of useful data delivered over the total energy consumption [58]. Finally, normalized throughput was defined as the number of bits successfully transmitted over the total number of generated bits within the same period.

HEH-BMAC vs. IEEE 802.15.6

Simulation results for the energy efficiency for both protocols, HEH-BMAC and IEEE 802.15.6, are shown in Figure 9.16. Energy efficiency for both schemes decreases as the number of low-priority BNs (L) grows. For L = 3, the energy efficiency for the IEEE 802.15.6 protocol reaches 2.36 Mb/J in Case 1 (low CW values) and 1.85 Mb/J in Case 2 (high CW values). This difference is observed because of the higher number of data collisions in Case 1, caused by the smaller values of CW bounds compared to Case 2.

In Figure 9.16, we can also see how the HEH-BMAC protocol initially (L = 3) has similar energy efficiency (1.89 Mb/J) in Case 1, but lower energy efficiency than IEEE 802.15.6 in Case 2. However, as L increases, our protocol gradually outperforms IEEE 802.15.6, regardless of the CW selection. For L = 18 (see Figure 9.16), HEH-BMAC achieves a gain of approximately 20% in energy efficiency with respect to the standard.

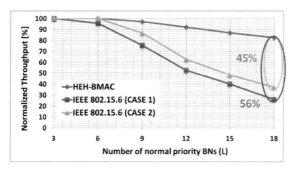

FIGURE 9.17: HEH-BMAC vs. IEEE 802.15.6: normalized throughputs

Figure 9.17 shows the normalized throughput for the two MAC protocols. We can observe how HEH-BMAC achieves higher throughput as the number of BNs increases, compared to the two cases of the IEEE 802.15.6. Looking at L = 18, we can see that the normalized throughput of the HEH-BMAC protocol overcomes the Case 1 and Case 2 of the IEEE 802.15.6 by 56% and 45%, respectively.

This performance enhancement is explained next. In IEEE 802.15.6, even though high priority nodes have a higher chance of accessing the medium, they may still suffer collisions with other normal or high priority BNs. On the contrary, in HEH-BMAC, the high priority BNs do not experience packet collisions, since the dynamic scheduling algorithm provides them with contention-free medium access. Packet collisions may only take place among the L normal priority BNs, but they are resolved in a dynamic way through the CP updating algorithm in the PC access.

To highlight the benefits of the dynamic scheduling algorithm of the HEH-BMAC, we compared the normalized throughput per BN of our protocol with Case 1 of the IEEE 802.15.6. Figure 9.18 shows the normalized throughput of the four high priority BNs and the average normalized throughput for L = 6, 12 and 18 BNs in normal priority for both HEH-BMAC and IEEE 802.15.6. Through the dynamic schedule algorithm in the HEH-BMAC, the normalized throughput of the high priority BNs is maintained to 100%. In IEEE 802.15.6, the normalized throughput of the high priority BNs is gradually reduced because of the increased number of collisions as L increases. The average normalized throughput of the normal priority BNs is decreased for both protocols. However, HEH-BMAC obtains better performance with respect to the standard.

9.7.1.6   Conclusions

HEH-BMAC is a novel hybrid polling MAC for WBANs powered by human energy harvesting. The novelty is that it is the first MAC protocol, to our knowledge, designed for WBANs in energy harvesting conditions. HEH-BMAC

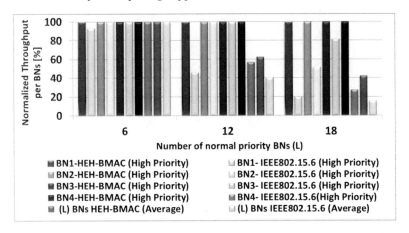

FIGURE 9.18: HEH-BMAC vs. IEEE 802.15.6: normalized throughput

adopts two modes of operation to provide priority differentiation to the sensor nodes and flexibility to the network. Moreover, HEH-BMAC is energy-aware, adapting its operation to changes in the energy supply of the nodes. Depending on the characteristics and needs of each node, the protocol assigns the most appropriate access mechanism (ID polling mode or PC access mode) to provide energy-aware priority-based scheduling.

Through a performance evaluation of our protocol for different energy harvesting rates, packet inter-arrival times, and network size, we observed that HEH-BMAC dynamically adapts its operation to potential changes in these parameters. Furthermore, we presented a performance comparison of the HEH-BMAC protocol to the IEEE 802.15.6 standard. Our protocol has been proven to outperform the IEEE 802.15.6 standard in terms of normalized throughput under the same conditions of energy harvesting. In addition, HEH-BMAC achieves higher energy efficiency when the number of nodes increases.

9.7.2   A Smart Algorithm for HEH-Powered BNs (PEH-QoS)

The idea of a WBAN that works in synergy with the human body is indeed promising. However, certain considerations must be taken into account to maintain an acceptable level of QoS in HEH-WBANs. Hortos [54, 59] studied the effect of energy harvesting on the QoS in real-time WSNs incorporating solar and wind energy harvesting techniques at the sensor nodes. The results revealed that the network's lifetime can be extended through the use of energy harvesting, but this may come at the cost of QoS degradation.

Prashanth et al. [60] performed stability measurements in the data queues of nodes powered by solar energy harvesting in indoor environments. This work showed that the stability of the data queue depends on the data arrival rate, the service time, and the waiting time. In addition, throughput degradation in the considered system was primarily due to increased waiting time of the data in the queue as a result of the energy variation.

Joseph et al. [61]] focused on increasing throughput and stabilizing the data queue through the development and implementation of optimal sleep-wake policies for energy harvesting sensor nodes. Their goals are fulfilled by allowing the node to sleep in some slots and drop some generated packets. Yang and Ulukus [62] developed optimal offline scheduling policies to minimize the time by which all packets are delivered to the sink in a single-user WSN under causality constraints on both data and energy arrivals. To achieve this goal, the transmission rate is adapted to the traffic load and available energy.

Adapting the BN operation to the variations of their energy level provides important benefits in energy harvesting systems. Khairnar and Mehta [63] proposed a transmission scheme for an energy harvesting node that achieves an average throughput close to the optimal achievable value. A node can only switch between a pre-specified set of discrete data rates and adjusts its power depending on its channel gain and its energy level. Murthy [64] developed an optimum transmission scheme focused on power management and data rate maximization for WSN-HEAP. Via this transmission scheme, applied to power-controlled nodes that transmit data using a fixed modulation and coding scheme, a specific data throughput can be guaranteed.

Ozel et al. [65] proposed adaptive transmission policies to maximize the average number of bits transmitted by a given node within a specific time. This energy management approach is able to adapt to the random energy arrivals and random channel fluctuations.

The papers cited above reveal how some authors address the problems introduced by energy harvesting in WSNs. In the current literature, we find various interesting approaches, e.g., the application of game theory to perform energy-aware transmission control in WSN-HEAP [66], and a power manager with a proportional integral derivative controller that enables the node to operate in energy neutral operation (ENO) state [67].

Several works address issues related to QoS guarantees in WBANs. However, the literature lacks papers that address the guarantee of QoS in HEH-WBANs. All the cited works show operation degradation at the nodes due to the energy harvesting, in packet transmission and also in the event detection. The time required to store the energy needed to detect or report (transmit) an event depends on the amount of collected energy in a given period. The main challenge to be faced by a BN powered by human energy harvesting is to maintain its functionality of detecting and transmitting correctly.

9.7.2.1 System Model

The BN architecture and the considered WBAN topology are depicted in Figure 9.19. We adopt a harvest-store-use architecture, and the WBAN is configured in star topology, assuming a single BN. To achieve a realistic model,

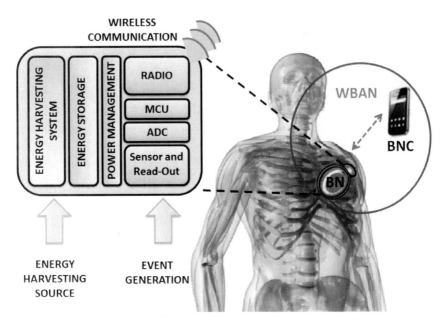

FIGURE 9.19: BN's architecture and WBAN topology: system model

we have chosen the characteristics based on current technological trends in low-power [68,69] and ultra-low power [70] wireless node designs for WBANs. We chose supercapacitors as energy storage units due to the benefits they provide to the system, e.g., they can charge and discharge at a faster rate than batteries.

It becomes clear that the collected energy exploitation is a key factor that will determine the smooth operation of a node. For this reason, we assume the energy management architecture proposed in [71] for biomedical devices using supercapacitors for energy storage. This energy management integrated circuit (EMIC) consists of a switched-capacitor DC-DC converter (which converts a source of direct current (DC) from one voltage level to another), a 4 nW bandgap voltage reference, a high-efficiency rectifier (to allow recharging of the capacitor bank), and a switch matrix and digital control circuitry (to govern the stacking and unstacking of the supercapacitors) [71].

Through the use of EMIC, more than 98% of the stored energy may be available for use. The transceiver is modeled as a 1.9 nJ/b 2.4 GHz multi-standard (Bluetooth Low Energy/Zigbee/IEEE802.15.6) transceiver for personal/body area networks [72], while the sensor and the reader are modeled based on a 30 W analog signal processor integrated circuit for biomedical signal monitoring [73]. These components were chosen because of their low power consumption and good reliability for WBANs.

In our model, the BNC is the sink responsible for setting up the WBAN and collecting all information transmitted by the BN. We assume that the

BNC is a smartphone that has higher processing capabilities than the BNs. The data communication between the BNC and BNs takes place via the ID polling access mode of the HEH-BMAC protocol.

We consider that our BNC has an external power supply, and the BN is able to harvest the energy available in the human body at a constant rate K_{EH}. In order to model a more realistic scenario, we adopted the same IAT_{BN} and packet size (l_{pkt}) as in [55] or event generation. We considered the BN power consumption to be divided into two main parts: the detection power consumption (P_{det}) and the transmission power consumption (P_{tx}). The term P_{det} includes the power consumption related to the correct BN operation (i.e., MCU, ADC, sensor, and read-out). P_{tx} includes the power consumption related to the duty cycle of the transceiver (i.e., data communication process).

9.7.2.2 Power-QoS Aware Management Algorithm

The proposed scheme, PEH-QoS, is executed at each BN, aiming to adapt the node's performance to its particular features, power supply, and QoS requirements. To that end, PEH-QoS combines three interconnected modules, illustrated in Fig. 9.20: (i) the power-EH aware management (PHAM), which calculates and manages the harvested energy, aiming to control the overall energy consumption of the system and keep the node in ENO state, (ii) the data queue aware control (DQAC), which manages the packet queue and ensures that only useful data are transmitted, by discarding any packets that have lost their clinical validity according to the delay and reliability requirements of the respective medical application, and (iii) the Packet Aggregator and Scheduling System (PASS), which uses the amount of power available for transmission (information from the PHAM module) and the amount of data stored in the queue (information from the DQAC module) to determine the maximum number of packets that can be transmitted in each data communication process. In the following sections, we describe the operation of each module in detail.

PHAM: Power-EH Aware Management:
The PHAM module, depicted in Figure 9.21, is responsible for the management of the harvested energy at the BN, which plays a critical role to the system's performance. PHAM uses the BN's power consumption information to distribute the harvested energy among the two key tasks of the node, namely, event detection and packet transmission. Typically, the energy consumed by the radio module of the BN is much higher compared to the energy required by the detector, i.e., $E_{tx} >> E_{det}$. Nevertheless, in WBANs, the detection capability of the BN can sometimes be more important than the transmission function, especially in the case of critical events.

Hence, the purpose of PHAM module is to efficiently manage the available energy in order to maximized the number of detected events and maintain the

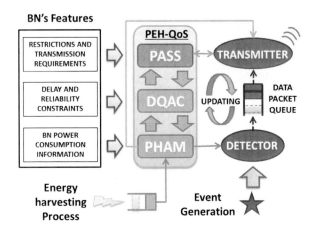

FIGURE 9.20: Illustration of the PEH-QoS operation

node operation in ENO state. The detector efficiency (D_{eff}) is given by:

$$D_{eff} = \frac{\text{Total number of events detected}}{\text{Total number of events occurred}} \qquad (9.1)$$

The level of energy stored in the battery (B_{level}) depends on the energy (E_{EH}) harvested from the human environment:

$$B_{level}(t) = B_{level}(t-1) + E_{EH} \qquad (9.2)$$

In turn, E_{EH} depends on the K_{EH} and the time (T_{EH}), which can be expressed as follows:

$$E_{EH} = \int_{0}^{T_{EH}} K_{EH} = K_{EH}T_{EH} \qquad \forall T_{EH} \in [0, \infty] \qquad (9.3)$$

In PHAM, first priority is to ensure the proper operation of the detector (i.e., $B_{level} \geq E_{det}$, as the transmissions take place only when the energy level reaches $B_{level} \geq E_{det} + E_{tx}$ (i.e., the available energy is sufficient for both transmission and detection). The amount of packets to be transmitted in each communication process, defined as D_{Load}, and the energy required for their transmission E_{tx} are calculated by the PASS module.

DQAC: Data Queue Aware Control:

Since the amount of harvested energy depends on the magnitude and availability of the energy harvesting source, data packets can remain stored for a long time before their transmission. This raises two major issues, the first related to the saturation of the data queue since the node has a finite storage capacity. Once the node has reached its maximum capacity of data storage,

---

ALGORITHM 1: *Power-EH Aware Management*

1: **if** $B_{level} \geq E_{det} + E_{Tx}$ **then**
2:     *Keep BN's detector* **ON**
3:     **if** $D_{Queue} \geq D_{Load}$ **then**
4:         *Keep BN's transceiver* **ON**
5:         *Make communication process of* D_{Load}
6:     **else**
7:         *Keep BN's transceiver* **SLEEP mode**
8: **elseif** $B_{level} < E_{det} + E_{Tx}$ **do**:
9:     **if** $B_{level} \geq E_{det}$ **then**
10:        *Keep BN's detector* **ON**
11:        *Keep BN's transceiver* **OFF**
12:    **else**
13:        *Keep BN's detector* **OFF**
14:        *Keep BN's transceiver* **OFF**
15: **end**
16: *Repeat step* 1

FIGURE 9.21: PHAM sub-module algorithm

it may not be able to store the next detected events and consequently these packets are lost.

The second problem is related to the loss of validity of the data stored because of the waiting time (e.g., in monitoring vital signals). DQAC is a sub-module designed to control the data queue and deal with these problems. The DQAC sub-module prevents the saturation (overflow) of the queue with unimportant data and allows all detected events to be stored.

DQAC performs the packet discard and update of the data queue (see Figure 9.22) using the information of maximum allowed end-to-end delay (Dly_{max}) and maximum storage capacity (SC_{max}). (Dly_{max}) depends on the BN's application requirements and (SC_{max}) is a physical restriction of the BN's hardware. DQAC constantly monitors the waiting times of each data packet (T_{pkt}) and the number of packets stored (DQ_{level}). The value of T_{pkt} must not exceed the maximum waiting time in the queue (T_{Qmax}); otherwise the data packet is deleted to release space in the queue. Deleted packets have lost their importance or have been deleted to make space for recent data packets. T_{Qmax} is calculated as:

$$T_{Qmax} = Dly_{max} - T_{TX} \qquad (9.4)$$

ALGORITHM 2: *Data Queue Aware Control*

1: if $DQ_{level} > 0$ **then**
2:    **Check** T_{pkt} *of all packets stored.*
3:     **if** $T_{pkt} \geq T_{Qmax}$
4:       **Delete** Data *packet.*
5:     **end**
6: **end**
7: if $DQ_{level} < SC_{max}$ **then**
8:    **if** event is detected **then**
9:       **Save** Data packet in the Queue
10:    **end**
11: **elseif** $DQ_{level} == SC_{max}$ **then**
12:    **if** event is detected **then**
13:       **Delete** packet with longer waiting time.
14:       **Save** new Data packet in the Queue
15:    **end**
16: **end**
17: *Repeat step* 1

FIGURE 9.22: DQAC sub-module algorithm

where T_{TX} is the time needed for the data communication, calculated in the PASS sub-module.

PASS: Packet Aggregation and Scheduling System:
In battery operated BNs, the data packets can be transmitted as soon as they are generated. In BNs operated by energy harvesting, data transmission can be realized when the node accumulates enough energy. PASS is a sub-module designed to optimize the data transmission in energy harvesting conditions (see Figure 9.23). The main objective of this algorithm is to send the maximum number of available data packets in every transmission. PASS uses the MAC protocol information for the calculation of E_{tx}.

Figure 9.24 shows the structure of the IEEE 802.15.6 data frame [30]. In our protocol, a variable number of D_{Load} packets are aggregated using the same control frames (depending on the applied protocol). D_{Load} depends on the BNs energy level (i.e., B_{level}) and the status of the data queue (i.e., DQ_{level}). An increase in the size of D_{Load} implies an increase in the required E_{tx}. Thus, the value of D_{Load} is indirectly adapted to the K_{EH} and the data arrival time of the BN.

ALGORITHM 3: *Packet Aggregator/Scheduling System*
1: **Calculate** E_{Tx} *required to send a single packet.*
2: **Check** B_{level} *and* DQ_{level} *status*
3: **if** $DQ_{level} = 1$ *and* $B_{level} \geq E_{Tx} + E_{det}$ **then**
4:   **Make** $D_{Load} = 1$
5:   **Perform** *data communication process.*
6: **elseif** $DQ_{level} > 1$ **then**
7:   **Determine** D_{Load}
8:   **Calculate** E_{Tx} *required to send* D_{Load}
9:   **if** $DQ_{level} \geq D_{Load}$ *and* $B_{level} \geq E_{Tx} + E_{det}$ **then**
10:     **Make** packet aggregation D_{Load}
11:     **Perform** *data communication process.*
12:   **elseif** $DQ_{level} < D_{Load}$ *and* $B_{level} \geq E_{Tx} + E_{det}$ **then**
13:     **Recalculate** D_{Load}
14:     **Calculate** E_{Tx} *required to send* D_{Load}
15:     **Make** packet aggregation D_{Load}
16:     **Perform** *data communication process.*
17:   **else** *wait to accumulate* $B_{level} \geq E_{Tx} + E_{det}$
18:     **Perform** *from Line* 13 *to Line* 16
19:   **end**
20: **end**
21: *Repeat step* 1

FIGURE 9.23: PASS sub-module algorithm

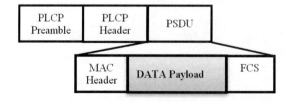

FIGURE 9.24: IEEE 802.15.6 data frame structure

9.7.2.3   Performance Evaluation

Simulation Consideration and Setup:
We developed an event-driven MATLAB simulator that implements our algorithm in a simple HEH-WBAN. We simulated a HEH-WBAN formed by one BNC and one BN. We assumed that the BNC has no energy shortage problems (it had an external power supply), while the BN was connected to an energy harvester that supplied energy to the node at a constant rate KEH. The BN stores the harvested energy in a rechargeable supercapacitor. As a BN, an ECG has been selected. Events detected by the ECG were converted into packets and then stored in a data buffer. The characteristics [55,68,72,73] and the QoS requirements [17] of the ECG are summarized in Table 9.3.

We assumed that the communication between the ECG and the BNC can take place directly without interference. The BNC provided medium access through the execution of the ID polling access of the HEH-BMAC. The network parameters were selected according to the IEEE 802.15.6 PHY-MAC specifications [30]. The system parameters used in the simulation are summarized in Table 9.4.

TABLE 9.3: ECG BN Characteristics

Data and	Packet arrival time		2 ms
	Data queue size		200 packets
Traffic Features	Packet size		12 bits
	Sensor READ-OUT and ADC		30 μW
	MCU		19.25 μW
Power		Reception	3.85 mW
Consumption	Transceiver	Transmission	4.6 mW
Distribution		Idle	0.712 mW
		Sleep	4 μW
QoS Requirements	Delay Constraint		< 250 ms
	Packet Loss Constraint		< 10%

TABLE 9.4: Simulation Parameters

Parameter	Value	Parameter	Value
Simulation Time	60 s	MAC Header	56 bits
p_{SIFS}	0.05 ms	FCS	16 bits
p_{CSMA} slot	0.125 ms	PLCP Preamble	90 bits
PLCP Tx rate	91.9 kb/s	PLCP Header	31 bits
Data Tx rate	485.7 kb/s	ACK	72 bits
Control Tx rate	121.4 kb/s	T_{POLL}	88 bits

Simulation Results:

Simulation results for the detection efficiency of ECG with and without PEH-QoS are shown in Figure 9.25. We can see how the PEH-QoS improves the performance of the ECG in detection efficiency. For very small values of K_{EH} (i.e., $K_{EH} < \frac{E_{det}}{T}$), the ECG without the PEH-QoS cannot detect any event. This is because the node is not aware of the available energy and it keeps trying to detect events although there is not enough energy for the detection; this wastes the collected energy. In this range, our scheme enhances the performance by 12.8% and 93.4% for $K_{EH} = 0.01$ mJ/S and $K_{EH} = 0.05$ mJ/S, respectively. In the case of $(K_{EH}) = 0.06$ mJ/S $(K_{EH} \geq \frac{E_{det}}{T})$, both systems achieve a detection efficiency of 100%, since there was sufficient energy for the detection of all the events.

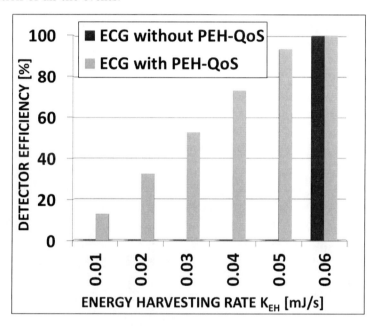

FIGURE 9.25: Detection efficiency

Figure 9.26 shows the behavior of the data queue for the two systems when $K_{EH} = 0.06$ mJ/S. The packets are continuously accumulated because the nodes do not have sufficient energy for transmission. In these conditions, our algorithm manages to stabilize the data queue, unlike the baseline scenario, where the data queue saturates. In Figure 9.26, the level of the queue is stabilized at 124 data packets (i.e., in this case $D_{Load} = 124$), where 100% of the stored information is valid, unlike the case where the algorithm is not applied and the queue is saturated with information that is no longer valid.

Figure 9.27 demonstrates that PEH-QoS is able to maintain the efficiency of data storage at 100%, which allows all sensory data to be stored until they are transmitted or lose their clinical validity. For the node without PEH-QoS

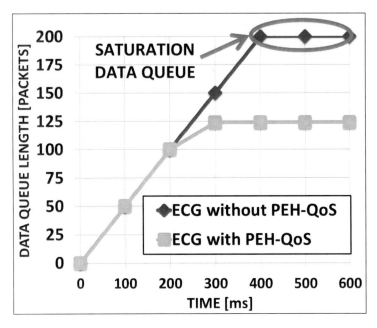

FIGURE 9.26: Data queue behavior

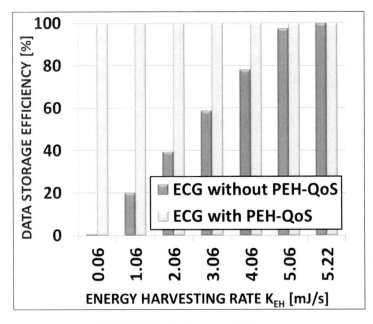

FIGURE 9.27: Storage efficiency

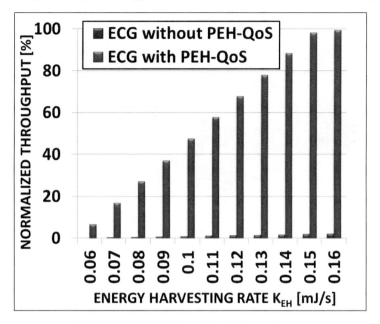

FIGURE 9.28: Normalized throughput

to reach good storage efficiency, a larger K_{EH} should be used but high values of K_{EH} cannot be achieved with current energy harvesting technologies.

Figure 9.28 presents the system throughput behavior for different values of K_{EH} for both schemes. Our scheme, with data aggregation $D_{Load} = 124$, significantly outperforms the baseline system. Our system in $K_{EH} = 0.16$ mJ/S reaches 100% of the normalized throughput while the other system is only reached 2.06% under the same conditions. This is justified by the fact that our scheme sends 124 data packets for transmission, unlike of the baseline that only transmits a single packet.

Figure 9.29 shows the variation in the normalized throughput in our scheme when applying different D_{Load} when $K_{EH} = 0.16$ mJ/S. Under the same testing conditions, Figure 9.30 and Figure 9.31 show the relationship between D_{Load} and the restrictions of reliability and delay, respectively. The value of $D_{Load} = 124$ data packets achieves the best delay value and reliability.

Figure 9.32 shows the energy needed to perform the transmission in both systems with and without PEH-QoS. In the case of ECG with PEH-QoS, for the transmission of a sequence of 124 data packets of 12 bits, the node spends only $E_{tx} = 24.3\mu J$. Conversely, in ECG without PEH-QoS, where no aggregation is applied (i.e., $L = 1$), the node needs $E_{tx} = 10.3\mu J$ for the transmission of a single data packet of 12 bits.

Figure 9.33 shows that our system is 50 times more energy efficient than the benchmark. In terms of packet loss (see Figure 9.34), our scheme fulfilled the reliability requirements of the application (i.e., maximum packet loss 10%),

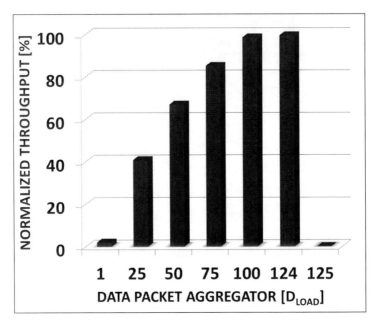

FIGURE 9.29: Normalized throughput vs. D_{Load}

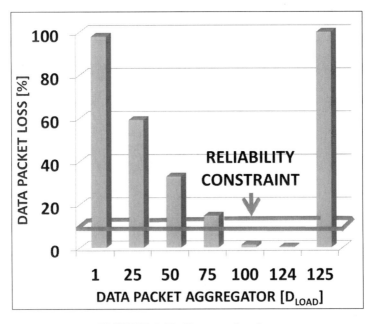

FIGURE 9.30: Data packet loss

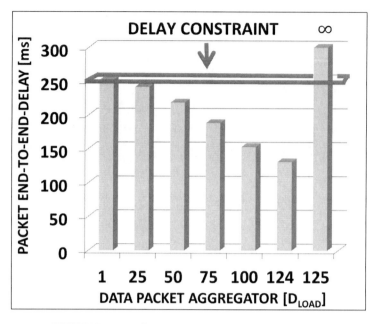

FIGURE 9.31: Average packet end-to-end delay

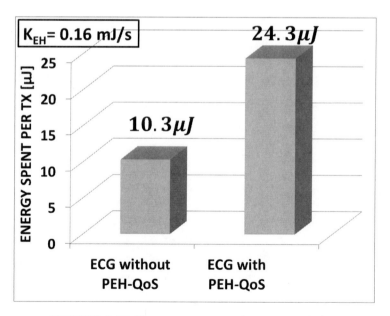

FIGURE 9.32: Energy spent per data transmission

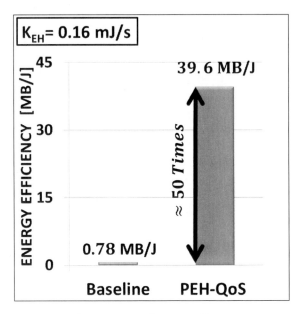

FIGURE 9.33: Energy efficiency

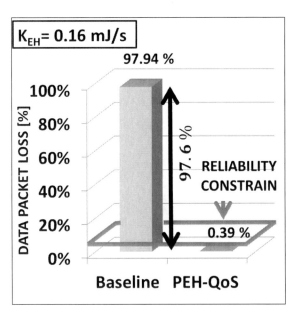

FIGURE 9.34: Packet loss

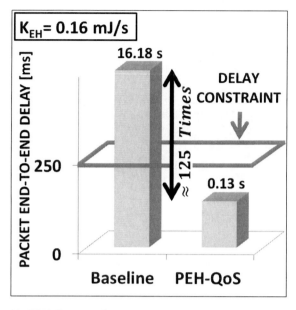

FIGURE 9.35: Average packet end-to-end delay

unlike the basic system which exceeded this threshold. Our system achieved 0.39% packet loss, while the baseline reached 97.94% (see Figure 9.34).

Finally, in Figure 9.35, we can see that the average packet end-to-end delay experienced in our system is 130 ms (maximum delay permitted is 250 ms), contrary to the baseline which obtained a much higher value (16.18 s).

9.7.2.4 Conclusions

PEH-QoS is a novel and highly efficient control scheme for BNs powered by energy harvesting. PEH-QoS consists of three sub-modules performing different tasks: PHAM maintains the node's operation in ENO state, DQAC controls and manages the data queue, keeping it updated with valid packet, and PASS optimizes data transmission, enhancing energy efficiency. The control scheme is designed to be executed on each node, under its specific energy harvesting conditions, achieving data transmission, while ensuring the proper functioning of the detection of events.

The most substantial conclusions can be summarized as follows:

- The application of our control scheme in the node achieved significant performance improvements in both data transmission and event detection function, attaining optimal management of the energy collected in the human environment. PEH-QoS can ensure that a BN can both detect medical events and transmit data packets efficiently.

- Extensive simulations have been conducted to evaluate the behavior of

PEH-QoS in a typical medical node under energy harvesting conditions. Our algorithm was proven to outperform the BN without PEH-QoS in terms of normalized throughput, energy efficiency, packet loss and average packet end-to-end delay.

BN with PEH-QoS has higher detection and storage efficiency than the baseline scheme under the same energy harvesting conditions. This feature enables BNs to detect and store all events, while avoiding the data buffer saturation.

9.8   Concluding Remarks on Soft Computing Potential

In this work, we provided energy-aware solutions, able to support and create autonomous, reliable, energy efficient HEH-WBANs. Our system was designed to operate in a star topology in which the BNC (smartphone or PDA) would provide medium access via execution of the HEH-BMAC protocol. Our research led to the development of two innovative approaches: (i) the HEH-BMAC protocol for sharing the medium and solving problems at the WBAN-level and (ii) the PEH-QoS control scheme responsible for solving problems in the BN-level and improving QoS.

To adapt to the challenging requirements of WBANs and time varying power supply through EH sources, a network requires flexibility, adaptability and enhanced intelligence. High complexity algorithms are not viable in medical applications because of the limited processing and power capabilities of the sensors. Soft computing technologies offer promising alternatives by providing low cost and low complexity adaptable solutions to meet the strict requirements of WBANs.

The European Centre for Soft Computing (ECSC) defines soft computing as "a set of computational techniques to solve problems by imitating nature's approaches" [74]. The main techniques of this discipline are fuzzy logic, neural networks, evolutionary computing, genetic programming and probabilistic reasoning.

Dhasian and Balasubramanian [75] presented a comprehensive summary of data aggregation techniques using soft computing in WSNs. They consider four types of soft computing techniques applied to data aggregation, namely fuzzy-based data aggregation, neural-based data aggregation, swarm-based data aggregation, and genetic-based data aggregation. Through the comparison of several network parameters, including energy consumption, cost, accuracy and security, they concluded that fuzzy and swarm artificial intelligent techniques can lead to accurate energy efficient solutions.

Lee et. al [76] suggested a particle swarm optimization classification system in the WBAN environment to analyze a large amount of blood pressure data efficiently, applying classification rules to improve the time performance.

The principles of soft computing can be adopted to further improve the two novel energy-harvesting-based mechanisms for WBANs proposed in this work, namely HEH-BMAC and PEH-QoS. Both schemes are dynamic, have low computational cost, and take into account various parameters (e.g. energy level, data inter arrival time, QoS requirements, etc.) and are perfect candidates to incorporate soft computing techniques.

HEH-BMAC introduces employs two access schemes (contention-free and contention-based) to prioritize nodes with different types of data and amounts of energy. In our initial approach, the priority assignment is deterministic. However, fuzzy logic principles could be employed to design a more flexible priority scheme, taking into account parameters such as data criticality, system load and channel quality. PEH-QoS contains an aggregation module (PASS) that selects the optimal number of aggregated packets based on the energy level of the node and the number of buffered packets. The PASS module can be enhanced by considering more sophisticated aggregation schemes, which may be based on fuzzy logic and swarm intelligent techniques [75].

The design of efficient schemes to exploit energy harvesting capabilities in WBANs is the key to autonomous and flexible mHealth platforms, able to revolutionize health care delivery. This chapter presented some first results, showing substantial improvements in performance, reliability and energy efficiency and stressing the need for further enhancements, possibly with the use of software computing techniques.

Acknowledgments

This work has been supported in part by AGAUR Project under Grant 2014-SGR-1551 and in part by the research project Kinoptim (324491).

Bibliography

[1] National Institute on Aging, National Institutes of Health and U.S. Department of Health and Human Services, Global health and aging, World Health Organization NIH Publication, 11-7737, pp. 1-32, October 2011.

[2] S. Mendis, P. Puska, and B. Norrving, Global atlas on cardiovascular disease prevention and control, World Health Organization, pp. 1-164, September 2011.

[3] B. Sanou, The world in 2013: ICT facts and figures, International Telecommunication Union, ICT Data and Statistics Division, February 2013.

[4] Global Observatory for eHealth series, mHealth new horizons for health through mobile technologies, World Health Organization Publication GOe Publication, vol. 3, pp. 1-112, June 2011.

[5] Vital Wave Consulting, mHealth for development: The opportunity of mobile technology for healthcare in the developing world, UN Foundation Vodafone Foundation Partnership, pp. 1-70, 2009.

[6] S. Simmons and T. Chan, Body area networks: driving mhealth growth, CSMG, pp. 1-5, March 2011.

[7] P. A. Gould and A. D. Krahn, Complications associated with implantable cardioverter-defibrillator replacement in response to device advisories, Journal of the American Medical Association, vol. 295, no. 16, pp. 529-551, April 2006.

[8] M. Parahuelva, Cardiovascular implantable cardioverter defibrillator-related complications: from implant to removal or replacement: a review, INTECH, pp. 101-116, September 2011.

[9] F. Wang and J. Liu, Duty-cycle-aware broadcast in wireless sensor networks, In Proceedings of IEEE INFOCOM, pp. 468-476, April 2009.

[10] C. Bachmann, M. Ashouei, V. Pop, M. Vidojkovic, H.D. Groot, and B. Gyselinckx, Low-power wireless sensor nodes for ubiquitous long-term biomedical signal monitoring, IEEE Communications Magazine, vol. 50, no. 1, pp. 20-27, January 2012.

[11] Silicon Laboratories Inc., How to design smart gas and water utility meters for the utmost in power efficiency, Silicon Labs Rev 1.1, 2013. http://www.silabs.com/Support%20Documents/TechnicalDocs/Low-Power-MCU-Metering.pdf.

[12] L. Hughes, X. Wang, and T. Chen, A review of protocol implementations and energy efficient cross-layer design for wireless body area networks, Journal of Sensors, vol. 12, no. 11, pp. 14730-14773, November 2012.

[13] S. Ullah, B. Shen, S. M. R. Islam, P. Khan, S. Saleem, and K. S. Kwak, A study of medium access control protocols for wireless body area networks, In Proceedings of IEEE GLOBECOM, pp. 13, April 2010.

[14] M.A. Ameen, A. Nessa, and K.S. Kwak, QoS issues with focus on wireless body area networks, In Proceedings of International Conference on Convergence and Hybrid Information Technology, pp. 801-807, November 2008.

[15] S. Drude, Requirements and application scenarios for body area networks, In Proceedings of IST Summit Mobile and Wireless Communications, pp. 1-5, July 2007.

[16] C. Kateretse, G-W Lee, and E-N Huh, A practical traffic scheduling scheme for differentiated services of healthcare systems on wireless sensor networks, Journal of Wireless Personal Communications, vol. 71, pp. 909-927, July 2013.

[17] X. Liang and I. Balasingham, A QoS-aware routing service framework for biomedical sensor networks, In Proceedings of International Symposium on Wireless Communication Systems, pp. 342-345, October 2007.

[18] S. Hassanpour, B. Asadi, Y. Vejdanparast, and P. Zargar, Improving reliability of routing in wireless body area sensor networks using genetic algorithm, in proceedings of ieee international conference on computer science and automation engineering, vol. 2, pp. 590-593, June 2011.

[19] G. R. Tsouri, A. Prieto, and N. Argade, On increasing network lifetime in body area networks using global routing with energy consumption balancing, Journal of Sensors, vol.12, pp. 13088-13108, September 2012.

[20] Md. A. Razzaque, C. S. Hong, and S. Lee, Data-centric multiobjective QoS-aware routing protocol for body sensor networks, Journal of Sensors, vol. 11, pp. 917-937, January 2011.

[21] B. Otal, L. Alonso, and C. Verikoukis, Novel QoS scheduling and energy-saving MAC protocol for body sensor networks optimization, In Proceedings of ICST International Conference on Body Area Networks, no. 27, pp. 1-4, March 2008.

[22] P.D. Mitcheson, E.M. Yeatman, G.K. Rao, A.S. Holmes and T.C. Green, Energy harvesting from human and machine motion for wireless electronic devices, Proceedings of the IEEE, vol. 96, no. 9, pp. 1457-1486, September 2008.

[23] W.K.-G. Seah, Z.A. Eu, and H.-P. Tan, Wireless sensor networks powered by ambient energy harvesting (WSN-HEAP) - Survey and challenges, International Conference on Proceedings of Wireless Communication, Vehicular Technology, Information Theory and Aerospace & Electronic Systems Technology, pp. 1-5, May 2009.

[24] S.A. Gopalan and J-T Park, Energy-efficient MAC protocols for wireless body area networks: Survey, International Congress on Ultra Modern Telecommunications and Control Systems and Workshops, pp. 739-744, October 2010.

[25] A. Rahim, N. Javaid, M. Aslam, Z. Rahman, U. Qasim, and Z.A. Khan, A comprehensive survey of mac protocols for wireless body area networks, BioSPAN with IEEE International Conference on Broadband and Wireless Computing, Communication and Applications, August 2012.

[26] J. Zhang, Y-X Liu, Ş.K. Özdemir, R-B Wu, F. Gao, X-B Wang, L. Yang, and F. Nori, Quantum internet using code division multiple access, Scientific Reports, vol. 3, no.2211, pp. 1-7, July 2013.

[27] N. Javaid, I. Israr, M. A. Khan, A. Javaid, S. H. Bouk, and Z. A. Khan, Analyzing medium access techniques in wireless body area networks, Research Journal of Applied Sciences, Engineering and Technology, pp. 1-17, April 2013.

[28] B. Zhen, R. Kohno, and H-B. Li, IEEE body area networks and medical implant communications, In Proceedings of the International Conference on Body Area Networks, pp. 3-6, March 2008.

[29] N. Javaid, S. Hayat, M. Shakir, M.A. Khan, S.H. Bouk, and Z.A. Khan, Energy efficient mac protocols in wireless body area sensor networks-a survey, Journal of Basic Applied Scientific Research, pp. 1-17, March 2013.

[30] 802.15. 62012 IEEE Stds. (2012). Standard for local and metropolitan area networks: Part 15.6: Wireless Body Area Networks.

[31] E. Kartsakli, A. Lalos, A. Antonopoulos, S. Tennina, M. Di Renzo, L. Alonso, and C. Verikoukis, A survey on M2M systems for mhealth: a wireless communications perspective, MDPI Sensors, vol. 14, no. 3, pp. 18009-18052, September 2014.

[32] N. F. Timmons and W.G. Scanlon, An adaptive energy efficient MAC protocol for the medical body area networks, In Proceedings of the 1st Wireless Communication, Vehicular Technology, Information Theory and Aerospace & Electronic Systems Technology, pp. 587-593, May 2009.

[33] G. Fang and E. Dutkiewicz, BodyMAC: Energy efficient TDMA-based MAC protocol for wireless body area networks, In Proceedings of the International Symposium on Communications and Information Technology, pp. 1455-1459, September 2009.

[34] H. Li and J. Tan, Heartbeat-driven medium-access control for body sensor networks, IEEE Transactions on Information Technology in Biomedicine, vol. 14, no. 1, pp. 44-51, January 2010.

[35] J. Khan and M. Yuce, Chapter 31: Wireless body area network (WBAN) for medical applications, new developments in biomedical engineering, pp. 591-627, 2010, InTech, DOI: 10.5772/7598. http://www.intechopen.com/books/new-developments-in-biomedical-engineering/wireless-body-area-network-wban-for-medical-applications.

[36] A. Boulis and Y. Tselishchev. Contention vs. polling: a study in body area networks MAC design, In Proceedings of the Fifth International Conference on BANs, pp. 98-104, September 2010.

[37] M.A. Ameen, N. Ullah, M. Chowdhury, S. Islam, and K. Kwak, A power efficient MAC protocol for wireless body area networks, EURASIP Journal on Wireless Communications and Networking, doi:10.1186/1687-1499-2012-33, February 2012.

[38] J. Polastre, J. Hill, and D. Culler, Versatile low power media access for wireless sensor networks, In *Proceedings of the ACM International Conference on Embedded Networked Sensor Systems (ACM SenSys)*, pp. 95-107, November 2004.

[39] M. Buettner, G.V. Yee, E. Anderson, and R. Han, X-MAC: a short preamble MAC protocol for duty-cycled wireless sensor networks, In Proceedings of the ACM International Conference on Embedded Networked Sensor Systems, pp. 307-320, November 2006.

[40] A. El-Hoiydi and J-D Decotignie, WiseMAC: an ultra low power MAC protocol for the downlink of infrastructure wireless sensor networks, In Proceeding of the 9th International Symposium on Computers and Communications, vol. 1, pp. 244-251, July 2004.

[41] Zigbee. Web page. 2013-12-12. http://www.zigbee.org

[42] B. Otal, L. Alonso, and C. Verikoukis, Efficient power management based on a distributed queuing mac for wireless sensor networks, In Proceedings of the IEEE Vehicular Technology Conference, pp. 105-109, April 2007.

[43] D. Yun, S-E. Yoo, D. Kim, and D. Kim, OD-MAC: An on-demand MAC protocol for body sensor networks based on IEEE 802.15.4, In Proceedings of the IEEE International Conference on Embedded and Real-Time Computing Systems and Applications, pp. 413-420, August 2008.

[44] E. Kartsakli, A. Antonopoulos, A. Lalos, S. Tennina, M. Di Renzo, L. Alonso, and C. Verikoukis, Reliable MAC design for ambient assisted living: moving the coordination to the cloud IEEE Communications Magazine, vol. 53, no. 1, pp. 78-86, January 2015.

[45] E. Kartsakli, A. Antonopoulos, L. Alonso, and C. Verikoukis, A cloud-assisted random linear network coding medium access control protocol for healthcare applications MDPI Sensors, vol. 14, no. 3, pp. 4806-4830, March 2014.

[46] A. Seyedi and B. Sikdar, Modeling and analysis of energy harvesting nodes in body sensor networks, In Proceedings of Medical Devices and Biosensors, pp. 175-178, June 2008.

[47] J. Ventura and K. Chowdhury, Markov modeling of energy harvesting body sensor networks, In Proceedings IEEE International Symposium on Personal Indoor and Mobile Radio Communications, pp. 2168-2172, September 2011.

[48] A. Seyedi and B. Sikdar, Energy efficient transmission strategies for body sensor networks with energy harvesting, IEEE Transactions on Communications, vol. 58, no. 7, pp. 2116-2126, July 2010.

[49] Y. He, W. Zhu, and L. Guan, Optimal resource allocation for pervasive health monitoring systems with body sensor networks, IEEE Transactions on Mobile Computing, vol. 10, no. 11, pp. 1558-1575, November 2011.

[50] Y. He, W. Zhu, and L. Guan, Optimal source rate allocation in body sensor networks with energy harvesting, In Proceedings of 2011 IEEE International Conference on Multimedia and Expo, pp. 1-6, July 2011.

[51] Z. A. Eu, H.-P. Tan, and W.K.G. Seah. Design and performance analysis of MAC schemes for wireless sensor networks powered by ambient energy harvesting, Journal of Ad Hoc Networks, vol. 9, no. 3, pp. 300-323, August 2011.

[52] Z.A. Eu and H-P Tan, Probabilistic polling for multi-hop energy harvesting wireless sensor networks, IEEE International Conference on Communications, pp. 271-275, June 2012.

[53] S. Ullah, B. Shen, S. M. R. Islam, P. Khan, S. Saleem, and K. S. Kwak, A study of medium access control protocols for wireless body area networks, In Proceedings of the IEEE Global Telecommunication Conference, pp. 1-13, April 2010.

[54] W.S. Hortos, Effects of energy harvesting on quality-of-service in transient wireless sensor networks, In Proceedings of Military Communications Conference, pp. 1-9, October 2012.

[55] J. Khan, M. Yuce, and F. Karami, Performance evaluation of a wireless body area sensor network for remote patient monitoring, In Proceedings of the 30th International Conference of the IEEE Engineering in Medicine and Biology Society, pp. 1266269, August 2008.

[56] 11073 ISO/IEEE Stds. (2008). Health informatics PoC medical device communication, Part 00101: guide–guidelines for the use of RF wireless technology (pp. 25-27). Piscataway: IEEE.

[57] B. Chen and D. Pompili, Transmission of patient vital signs using wireless body area networks, Journal of Mobile Networks and Application, vol. 16, no. 6, pp. 663-682, December 2011.

[58] V. Rodoplu and T.H. Meng, Bits-per-joule capacity of energy-limited wireless networks, IEEE Transactions on Wireless Communications, vol. 6, no. 3 pp. 85765, may 2007.

[59] W. S. Hortos, Effects of energy harvesting on quality-of-service in real-time, wireless sensor networks, In Proceedings of, Wireless Sensing, Localization, and Processing VII, vol. 8404 840407, p.p. 1-12, May 2012.

[60] V. Prashanth, T.V. Prabhakar, K. Prakruthi, and H.S. Jamadagni, Queue stability measurements for energy harvesting sensor nodes, National Conference on Communications, pp. 1-5, February 2012.

[61] V. Joseph, V. Sharma, and U. Mukherji, Optimal sleep-wake policies for an energy harvesting sensor node, In Proceedings of IEEE International Conference on Communications, pp. 1-6, 14-18 June 2009.

[62] J. Yang and S. Ulukus, Optimal packet scheduling in an energy harvesting communication system, IEEE Transactions on Communications, vol. 60, no. 1, pp.220-230, January 2012.

[63] P.S. Khairnar and N.B. Mehta, Power and discrete rate adaptation for energy harvesting wireless nodes, In Proceedings of IEEE International Conference on Communications, pp. 1-5, June 2011.

[64] C.R. Murthy, Power management and data rate maximization in wireless energy harvesting sensors, In Proceedings of IEEE 19th International Symposium on Personal, Indoor and Mobile Radio Communications, pp. 1-5, September 2008.

[65] O. Ozel, K. Tutuncuoglu, J. Yang, S. Ulukus, and A. Yener, Adaptive transmission policies for energy harvesting wireless nodes in fading channels, In Proceedings of 45th Annual Conference on Information Sciences and Systems, pp. 1-6, March 2011.

[66] F-Y. Tsuo, H. Tan, Y. H. Chew, and H-Y Wei, Energy-aware transmission control for wireless sensor networks powered by ambient energy harvesting: a game-theoretic approach, In Proceedings of IEEE International Conference on Communications, pp. 1-5, June 2011.

[67] T. N. Le, O. Sentieys, O. Berder, A. Pegatoquet, and C. Belleudy, Power manager with PID controller in energy harvesting wireless sensor networks, In Proceedings IEEE International Conference on Green Computing and Communications, pp. 668-670, November 2012.

[68] C. Bachmann, M. Ashouei, V. Pop, M. Vidojkovic, H.D. Groot, and B. Gyselinckx, Low-power wireless sensor nodes for ubiquitous long-term biomedical signal monitoring, IEEE Communications Magazine, vol. 50, no. 1, pp.20-27, January 2012.

[69] B. Otis, J. Holleman, Y. -T. Liao, J. Pandey, S. Rai, Y. Su, and D. Yeager, Low power IC design for energy harvesting wireless biosensors, In Proceedings of IEEE Radio and Wireless Symposium, pp. 5-8, January 2009.

[70] L. Huang, M. Ashouei, R. Yazicioglu, J. Penders, R. Vullers, G. Dolmans, P. Merken, J. Huisken, H. de Groot, C. Van Hoof, and B. Gyselinckx, Ultra-low power sensor design for wireless body area networks - challenges, potential solutions, and applications, International Journal of Digital Content Technology and its Applications, vol. 3, no. 3, pp. 136-148, September 2009.

[71] W. Sanchez, C. Sodini, and J.L. Dawson, An energy management IC for bio-implants using ultracapacitors for energy storage, In Proceedings of IEEE Symposium on VLSI Circuits, pp. 63-64, June 2010.

[72] Y.-H. Liu, X. Huang, M. Vidojkovic, A. Ba, P. Harpe, G. Dolmans, and H. de Groot, "A 1.9nJ/b 2.4GHz multistandard (Bluetooth Low Energy/Zigbee/IEEE802.15.6) transceiver for personal/body-area networks," In Proceedings of IEEE International Solid-State Circuits Conference Digest of Technical Papers, pp. 446-447, February 2013.

[73] R. F. Yazicioglu, S. Kim, T. Torfs, P. Merken, and C. Van Hoof, A 30 analog signal processor ASIC for biomedical signal monitoring, In Proceedings of IEEE International Digest of Technical Papers Solid-State Circuits Conference, pp. 124-125, February 2010.

[74] European Centre for Soft Computing, What is soft computing?, http://www.softcomputing.es/metaspace/portal/3/73

[75] H.R. Dhasian and P. Balasubramanian, Survey of data aggregation techniques using soft computing in wireless sensor networks, IET Information Security, vol. 7, no. 4, pp. 336-342, December 2013.

[76] M. Lee, L. Tian, and O. Myagmar, Design of a particle swarm optimization-based classification system in a WBAN environment, International Journal of Advances in Computer Science and Technology, vol. 2, no. 8, pp. 24-27, August, 2013.

Chapter 10

Complex Network Entropy

Priti Singh

Indian Institute of Space Science and Technology

Abhishek Chakraborty

Indian Institute of Space Science and Technology

B. S. Manoj

Indian Institute of Space Science and Technology

10.1   Introduction

With the growth in technology, networks are continuously evolving and increasing in size and complexity. Internet and World Wide Web networks provide a platform where users connect to other users for an extensive exchange of information. Such networks are becoming more complex with their rapid growth. Additions of more nodes and links have increased the complexity further.

Various network models are suggested to analyze the growth patterns and complex nature of the networks [2]. Average path length (APL) and clustering coefficient are important parameters describing different complex network models. APL is defined as the mean distance between any pair of nodes in the network and clustering coefficient refers to the fraction of the pairs of neighbors of a node that are directly connected to each other.

Erdős and Rényi (ER) first proposed the random network model [1]. The node linkage in random networks follow a Poisson distribution and every pair of nodes in random networks is connected independently with the same probability. A random network does not show clustering and has smaller APL values between two nodes [2]. Real-world complex networks do not entirely follow random network models. Real-world complex networks, have the properties of both random and regular networks and incorporate small-world and scale-free features [2].

Watts and Strogatz (WS) developed the small-world network model which is transformed from a regular network model by rewiring links [3] in such a way that the APL of the WS small-world network decreases and the clustering coefficient increases [2]. The other important discovery that helped in the understanding of complex networks is the scale-free model [2].

Barabási and Albert [4] proposed a model where most of the nodes in the network have only a few edges and a small fraction of nodes have large numbers of edges. The scale-free model is more inhomogeneous than the random network and the small-world models which are homogeneous in degree distribution [2]. The scale-free model shares most of the characteristics of real-world networks [2]. To understand the properties of a complex network such as complexity, robustness, and growth pattern, it is important to understand the variations of those properties with the parameters of a network. Network parameters include degree of a node, link or edges, connectivity of a node, centrality of a node, clusters, cycles, and loops in social networks, topology configurations, and adjacency matrices. The variation of network properties among complex network models was studied using entropy.

Entropy of a network considers network parameters to create a probability metric to analyze complex networks. The metric assigns probability to each network state and an average amount of information is collected from all the states to reveal the overall network behavior. The state defines the behavior of each node or link in a network. The entropy definition [5] given by Shannon describes entropy as the measure of the average amount of information that can be extracted from a message transmitted from a source. According to information theory, the amount of information retrieved from a message that has a higher probability of occurrence is less than the amount of information retrieved from a message that has a lower probability of occurrence. Therefore, self information of a message $I(\cdot)$ is inversely proportional to the probability of occurrence $p(\cdot)$ and is defined as a logarithmic function of $p(\cdot)$ as $I(\cdot) = \log(\frac{1}{p(\cdot)})$. Thus, the average amount of information given by entropy $H(\cdot)$ is defined as $H(\cdot) = -\sum p(\cdot) \log(p(\cdot))$. Using Shannon's entropy definition, we see that equiprobable events correspond to maximum uncertainty in the system.

Shannon's definition of entropy can be extended to analyze network entropy by taking probability of occurrence $p(\cdot)$ of a message from a source as the probability of occurrence $p(\cdot)$ of a parameter. Thus, an average amount of information about a network is obtained by the probability of occurrence of a network parameter. The entropy measure is helpful in analyzing properties such as heterogeneity, connectivity, centrality, formation of clusters, cycles in a network, and the evolution of a network over time. Entropy is also a measure of complexity of a network ensemble [6].

The objective of this chapter is to summarize some of the existing entropy definitions used to explain properties of complex networks and a new entropy measure is proposed based on the variations in link lengths in a network. In Section 10.2, each entropy definition is used to analyze the variation in the corresponding network parameter to explain network properties. In Section 10.3, a new definition is proposed by which entropy of complex networks can be calculated by taking state to be the length of a link in the network. In Section 10.4, a comparison is made among all the entropy definitions of Section 10.2 and computational time complexity in estimating entropy is calculated for each case. Finally, Section 10.5 concludes this chapter.

10.2 Review of Existing Strategies

Some existing entropy definitions based on different parameters are discussed here. Each entropy definition explains certain unique features of complex networks depending on the parameter. Entropy of a network is measured based on nodal degree [7]– [9], topology configuration [10], connectivity [11], centrality [12], clustering [13], number of cycles or loops [15], automorphism partition [16], and navigation [17]. Entropy based on nodal degree is categorized as entropy of degree sequence [7] and entropy of degree distribution [8].

10.2.1   Entropy Based on Nodal Degree

Degree method to determine entropy is useful for measuring heterogeneity which in turn determines the robustness of a network. Heterogeneity is determined by observing nodal degrees in a network. If all the nodes have equal degree, the network is homogeneous. A slight deviation of a few nodes introduces heterogeneity in the network. The extent of deviation from homogeneous nature is determined by calculating entropy and thus provides an effective way to understand the vulnerability of a network in case of possible link or node failure. Entropy based on degree is defined as:

$$H = -\sum_k p(k) \ln p(k) \tag{10.1}$$

where $p(k)$ refers to the probability of fraction of nodes having degree k or the degree of the kth node.

Based on how $p(k)$ can be estimated, nodal degree entropy can be classified as: (a) entropy of degree sequence or (b) entropy of degree distribution.

10.2.1.1   Entropy of Degree Sequence

In entropy of degree sequence, the probability $p(k)$ of the kth node is directly proportional to its number of links. $p(k)$ is defined as:

$$p(k) = \frac{d(k)}{\sum\limits_{k=1}^{N} d(k)} \tag{10.2}$$

where $d(k)$ is degree of kth node in an N node network. Entropy of degree sequence is useful for measuring heterogeneity in scale-free networks. Entropy is maximum for homogeneous or regular networks, where $p(k) = \frac{1}{N}$; therefore, $H_{max} = \ln N$. Entropy is minimum for the most heterogeneous networks where $H_{min} = \frac{\ln 4(N-1)}{2}$ [7].

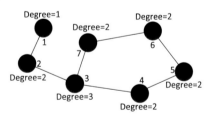

FIGURE 10.1: Degree sequence-based entropy

In Figure 10.1, entropy of the network is $H = -\{\frac{1}{14} \ln \frac{1}{14} + 5 * \frac{2}{14} \ln \frac{2}{14} + \frac{3}{14} \ln \frac{3}{14}\} = 1.90$, whereas, if degree is equally distributed among all the nodes in the network, we achieve maximum entropy which is $H_{max} = 1.95$ using Equation (10.1) and Equation (10.2). The deviation of the network from the maximum entropy tells us the amount of heterogeneity present.

Heterogeneity in scale-free networks is measured by evaluating normalized network structure entropy ($NNSE$) as described in Equation (10.3) [7].

$$NNSE = \frac{H - H_{min}}{H_{max} - H_{min}}; 0 \leqslant NNSE \leqslant 1 \qquad (10.3)$$

When $NNSE = 1$, a network is the most homogeneous as entropy is maximum. $NNSE$ value decreases as network becomes heterogeneous and entropy decreases from its maximum value.

Scale-free networks with varying sizes are compared on the basis of $NNSE$ value. These networks follow a power law distribution of the form $p(k) = ck^{-\alpha}$, where, $p(k)$ is the fraction of nodes having degree k. Degrees of all nodes are arranged in decreasing order to obtain a degree sequence. With the help of probability distribution of scale-free networks, a relation is obtained between a particular degree d and the rank of that degree in the degree sequence.

Suppose $d = f(r)$ is the relationship between degree and rank in the sequence. The rank function obtained is then substituted in place of degree d in Equation (10.2) to obtain the expression of entropy for scale-free networks. The obtained entropy function is dependent on the scaling exponent α and minimum degree of the network. By keeping N and minimum degree fixed, the entropy of scale-free networks is minimum when $\alpha \approx 1.7$ at which scale-free networks reach maximum heterogeneity. Also for $\alpha > 2$, entropy of scale-free networks becomes independent of the minimum degree of the network [7].

10.2.1.2 Entropy of Degree Distribution

Entropy of degree distribution [8] is useful for understanding heterogeneity in scale-free networks; $p(k)$ is expressed as follows:

$$p(k) = ck^{-\alpha} \qquad (10.4)$$

where $p(k)$ is the probability of the number of nodes having degree k, α is the scaling factor and c is a constant. Entropy of degree distribution is used to measure the robustness of the network. Since scale-free networks follow the power law degree distribution probability function as given in Equation (10.4), a large fraction of nodes have smaller degrees as compared to a small fraction of nodes having greater degrees, thus introducing heterogeneity in scale-free networks. Heterogeneity in scale-free networks makes them more robust to random attacks.

The optimal entropy of the network is achieved using the entropy of degree distribution method and percolation theory [18] to study robustness by random node removal. Using percolation theory, the threshold value at which the network disintegrates on randomly removing nodes from the network is determined. A relationship has been derived between entropy of degree distribution and network threshold value [8] under random node removal. For different average degree k, network threshold value increases with entropy of degree distribution. Therefore, entropy of degree distribution is an effective way to measure robustness of a network.

Degree-based entropy analysis is also used to explain evolution of networks over time as discussed later in this chapter.

10.2.2 Entropy Based on Topology Configuration

Entropy based on topology configuration [10] measures entropy variations across an ensemble of network configurations for the Erdős Rényi (ER) network model. Different configurations are obtained by changing the network size N and link probability p. If N is the total number of nodes in the network, the maximum number of links in a network is $M = \frac{N(N-1)}{2}$. Entropy measure based on topology configuration is defined as:

$$H(N, p) = - \sum_{k=0}^{M} C_k^M p(k) \log_2 p(k) \tag{10.5}$$

where M is the maximum number of links present in an N node network, k is the actual number of links present in the configuration, and $p(k)$ is the probability of the kth configuration state. $p(k)$ is defined as:

$$p(k) = p^k (1 - p)^{M-k} \tag{10.6}$$

On further simplifying Equation (10.5) for an ER random network model where $M \to \infty$ as $N \to \infty$, topology entropy is obtained as:

$$H(N, p) = -M[p \log_2 p + (1 - p) \log_2(1 - p)] \tag{10.7}$$

Entropy values vary significantly across different network configurations. For a particular value of N, maximum entropy is achieved when $p = 0.5$ and minimum entropy is obtained for $p = 0$ and $p = 1$ using Equation (10.6). The maximum entropy value increases as the number of nodes in the topology increases. Topology of a network plays a crucial role in determining the behavior or growth pattern based on the change in configuration of a complex network. Thus, topology entropy explains the dynamics of a network as the entropy value varies significantly across different random network models with varying numbers of nodes.

Topology-based entropy has many applications in mobile ad-hoc networks (MANETs). A MANET is a wireless network where every node is free to move independently in any direction giving rise to topologies which are dynamically changing. In [19], topological entropy is measured in a MANET, where a mobility metric-based entropy is calculated to ensure node connectivity.

Since nodes are continuously moving, mobility should be a measure of topological change. Here node mobility is described based on the random walk mobility model (RWMM) following a Markov process. Uncertainty in node movement creates uncertainty in the topology and is called topological uncertainty.

Connectivity in MANETs is determined by collecting past information about topological changes using finite L-block approximation [19] to create a mobility metric to predict future changes in topology. L-block represents the L consecutive topological changing states of the network. Thus, the metric determines the minimum information or overhead required by the connectivity service to measure node mobility.

In a way, the connectivity service is bounded by node mobility factors for estimating entropy based on topological change. By correctly identifying the topological change using the proposed mobility metric, node connectivity is ensured and thus allows packets to be routed through these nodes from source to destination. Therefore, identifying topological uncertainty is important for correct routing of packets in a network where the nodes are mobile.

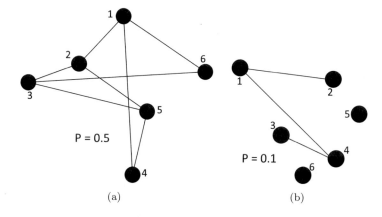

FIGURE 10.2: ER random network configurations: (a) network with link probability $p = 0.5$ and (b) network with link probability $p = 0.1$

In Figure 10.2(a) and Equation (10.7), $N = 6$, therefore, $M = \frac{N(N-1)}{2} = 15$. Entropy of the ER network in Figure 10.2(a) is maximum with $p = 0.5$ and is equal to 15, whereas entropy decreases in Figure 10.2(b) with $p = 0.1$ and is equal to 7.035 using Equation (10.7). Variation in the growth pattern of links based upon the link probability of a particular configuration is described in the example keeping N fixed as shown in Figure 10.2. Entropy of the network is less in a network with smaller link probability and entropy is maximum in the case of maximum uncertainty in link connection. Moreover, when the growth involves nodes, topology entropy increases as the number of nodes increases.

10.2.3 Network Connectivity Entropy

Network connectivity is determined by the minimal number of components required for the network to be connected. According to degree-based entropy definition as described in Section 10.2.1, probability of isolated nodes is zero since they have no degrees, therefore, these nodes do not contribute to network entropy. Network connectivity entropy [11] takes into account isolated nodes in the network. Thus, connectivity entropy is an effective measure to compare reliability between disconnected networks. Comparing reliability is important to indicate which network is more effective for exchange of information from source to the correct destination in an appropriate time. Network connectivity entropy is given as follows:

$$H = -\sum_{k=1}^{n} p(k) \ln p(k) \tag{10.8}$$

where N is total number of nodes in a network graph G and n is the number of connectivity sub-graphs in a network such that $(1 \leqslant n \leqslant N)$. $p(k)$ is the probability of the kth connectivity sub-graph and depends on the number of nodes present in the kth sub-graph represented by n_k. Number of nodes present in a connectivity sub-graph determines the importance of the sub-graph. $p(k)$ is given as:

$$p(k) = \frac{n_k}{N} \tag{10.9}$$

Breakdown of links or edges in a network can create sub-graphs or isolated nodes. Thus, calculating entropy based on degree is not an effective measure of information of all the nodes present in a network. Network connectivity entropy overcomes such limitation and measures the entropy of disconnected networks and calculates connectivity reliability as given in the following equation.

$$R = \frac{H_{max} - H}{H_{max} - H_{min}} = \frac{\ln n - H}{\ln n} \tag{10.10}$$

According to this equation, maximum connectivity entropy is achieved when all nodes are isolated or when all links are removed. Therefore, $H_{max} = \ln n$, and reliability is minimum ($R = 0$) for $H = H_{max}$. $H_{min} = 0$ is the minimum entropy of a connected graph and the connectivity reliability is maximum ($R = 1$) when $H = H_{min}$.

In Figure 10.3(a), $H = 0$, $R = 1$. In Figure 10.3(b), $H = -\{\frac{6}{12} \ln \frac{6}{12} + \frac{2}{12} \ln \frac{2}{12} + \frac{4}{12} \ln \frac{4}{12}\} = 1.01$, $R = 0.59$. In Figure 10.3(c), $H = -\{\frac{7}{12} ln \frac{7}{12} + \frac{1}{12} ln \frac{1}{12} + \frac{4}{12} ln \frac{4}{12}\} = 0.887$, $R = 0.64$. The entropy of each network configuration is calculated using Equation (10.8) and Equation (10.9). Reliability is calculated using Equation (10.10), and is maximum for a connected network as shown in Figure 10.3(a). Thus, in disconnected networks as shown in Figure 10.3(b) and Figure 10.3(c), connectivity reliability decreases from the maximum and is easily compared to determine a more reliable network.

Network connectivity entropy measures the degree of connectivity of each node or link when the removal of links creates sub-graphs or may even create isolated nodes. A change in the network connectivity entropy on the creation of an isolated node is a measure of that node's connectivity in the network. Removal of a link from a lower density network causes greater change in connectivity entropy as compared to a highly dense network.

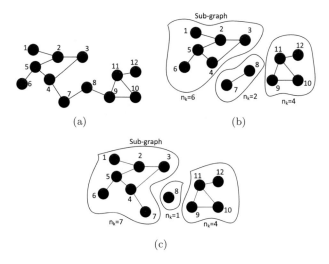

FIGURE 10.3: Connectivity entropy: (a) connected network, (b) disconnected network, and (c) disconnected network with isolated node

If the number of links in a partially or fully connected network are reduced, connectivity entropy increases or may even remain the same when no sub-graphs are created on the removal of a link and the maxima is achieved when all nodes become isolated. In other words, if we start adding links to a network with all isolated nodes, connectivity entropy decreases or may remain the same and the minima is achieved when the network becomes connected.

10.2.4 Network Centrality Entropy

Centrality refers to the importance of a node in the network. Network central-ity entropy [12] measures average information of the most important node. It measures heterogeneity in complex networks with respect to the number of shortest paths from a node in a bidirectional network. Centrality entropy is given as:

$$H = -\sum_{i=1}^{n} p(k_i) \log_2 p(k_i) \tag{10.11}$$

where $p(k_i)$ is given as in Equation (10.12).

$$p(k_i) = \frac{spaths(k_i)}{spaths(k_1, \ k_2, \ k_3, \ldots, \ k_M)} \tag{10.12}$$

where k_i refers to the ith node, $spaths(k_i)$ represents number of shortest paths from k_i to all other nodes in the network graph and $spaths(k_1, \ k_2, \ k_3, \ldots, \ k_M)$ represents total number of shortest paths in a network graph.

In Figure 10.4, each node is separated from its adjacent node by a unit

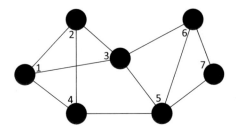

FIGURE 10.4: Centrality entropy

distance and node 3 acts as the most central node as it has the highest central-ity value given in Table 10.1. Entropy of the network is equal to 1.456 using Equation (10.11). Maximum centrality entropy is obtained in a fully connected network since each node can be reached by other nodes by the same number of shortest paths; therefore, all the nodes share equal importance in the net-work. The maximum entropy $H_{max} = \log_2 N = 1.946$ is obtained when all the 7 nodes are equally important.

Centrality entropy is a measure of path distribution commonly used in social networks to quantify the influence of a specific node over other nodes in a network. Centrality of a node depends on location which determines the number of its shortest paths. The more shortest paths through that node, the higher the probability of occurrence of that node. Uncertainty of that node decreases with higher probability, and such nodes have a greater impact on the network.

If all the nodes are equally important as in the case of fully connected net-works, removal of any node causes same effect on entropy. In a partially con-nected network with heterogeneous path distributions, if any node connected to the network is such that its removal considerably reduces the shortest paths of other nodes in the network, the centrality entropy decreases to a greater extent. Therefore, centrality entropy is maximum for a fully connected net-work and decreases for a partially connected network with heterogeneous path distribution. The basic idea is to find nodes in a network which will produce largest change in centrality entropy when removed.

TABLE 10.1: Centrality Values for Network Shown in Figure 10.4

Node Number	Centrality Value
1	0.0417
2	0.0417
3	0.3988
4	0.1190
5	0.2619
6	0.0833
7	0.0536

10.2.5 Clustering Entropy

Clustering is a technique by which nodes group to form a cluster so that nodes in the same cluster share similar properties as compared to nodes in other clusters. Clustering entropy [13] measures the degree of similarity among nodes in the cluster. An optimal cluster is found using an entropy-based clustering algorithm. To obtain the optimal cluster, a random seed node is chosen from a network and a cluster is formed using the chosen node and all its neighbors. Next, entropy of the cluster is calculated and the process is iteratively repeated until and unless a cluster with minimal entropy is obtained each time by the removal and addition of nodes to the cluster boundary. An optimal cluster has minimum cluster entropy. The whole process is repeatedly performed until all the nodes have been clustered.

To describe the entropy of a cluster, a cluster G' containing a random seed node and all its neighbors is selected from a network G. The inner links have links from node k to other nodes in G' and outer links have links from a node k to other nodes in G excluding the nodes in G'. Clustering entropy is given as follows:

$$H(G) = \sum_k H(k) \tag{10.13}$$

where $H(G)$ refers to the network clustering entropy and $H(k)$ refers to the nodal entropy. $H(k)$ is defined as:

$$H(k) = -p_i(k) \log_2 p_i(k) - p_o(k) \log_2 p_o(k) \tag{10.14}$$

where $p_i(k)$ refers to the probability that a node k has (inner) links only with other nodes inside the cluster. $p_o(k)$ refers to the probability that a node k has (outer) links with nodes not in the cluster. $p_i(k)$ and $p_o(k)$ is given below.

$$p_i(k) = \frac{n}{N(k)} \tag{10.15}$$

$$p_o(k) = 1 - p_i(k) \tag{10.16}$$

In Equation (10.15), n is the number of inner links of a node k and $N(k)$ refers to the total number of neighboring nodes of the kth node in a network G.

Figure 10.5(a) illustrates a cluster G' containing vertices $\{1, 2, 3, 4, 5\}$, where, vertices $\{1, 2, 3, 4\}$ have all inner links. Therefore, $p_i(1) = p_i(2) = p_i(3) = p_i(4) = 1$, which makes vertex entropy of nodes $\{1, 2, 3, 4\}$ to be zero using Equation (10.10). Vertex 5 has links inside and outside the cluster, thus $p_i(5) = 0.80$, $p_o(5) = 0.20$ using Equation (10.11) and Equation (10.12). Using Equation (10.9), we can obtain $H(5) = 0.722$, $H(6) = 1$. Therefore, $H(G) = H(5) + H(6) = 1.722$ whereas entropy of network graph in Figure 10.5(b) is $H(G) = H(4) + H(5) = 1.918$. Thus, network entropy in Figure 10.5(b) increases due to lower intra-connections among the nodes in the cluster.

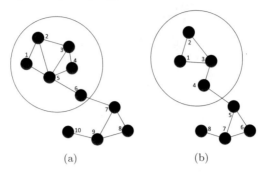

(a)   (b)

FIGURE 10.5: Cluster entropy [13]

Clustering entropy is used to study cluster quality of a network. High cluster quality is achieved by determining an optimal cluster with minimum entropy which depends on the types of intra-connections among the nodes belonging to the same cluster and interconnections with nodes outside the cluster. Strong intra-connections between the nodes in the cluster lead to higher cluster quality and vice versa. Thus, clustering entropy is a qualitative measure for analyzing communities in social networks.

10.2.6   Entropy Based on Number of Cycles and Loops

Cycles and loops are integral parts of social networks. Friendship networks, for example, employ loops to study how many neighbors of a node are also neighbors of each other. Degree-based entropy cannot be used to classify such networks [14]. Cycles provide feedback paths for such networks [15]. The more cycles in a network, the more connected it is. Cyclic entropy measures the storing ability of cycles in a network. Entropy variation has been studied across different types of networks (random, scale-free, and small-world) to calculate and compare minimum, maximum, and optimum entropies. Entropy measurement based on number of cycles is calculated as:

$$H = -\sum_{k} p(k) \log p(k) \qquad (10.17)$$

where $p(k)$ is the probability of finding a cycle of length k and the state of the network is taken as number of cycles of length k in the network. Cyclic entropy in Figure 10.6 is equal to 1.37 using Equation (10.17).

In [15], cyclic-based entropy analysis is compared with degree-based entropy analysis for a real-world complex network, random network, small-world network, and scale-free network. Cyclic entropy was calculated for each network with varying sizes. The results showed that the plot of both real and random networks had smaller variation with respect to the network size but the cyclic entropy values for a real network were closer to small-world net-

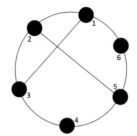

FIGURE 10.6: Partially connected cyclic network

work values when the network was small. The small-world network had the minimum cyclic entropy if the network size was 150. Maximum entropy was found in the scale-free networks where the cyclic entropy varied significantly with the network size. Thus, the scale-free model is said to be more robust because of its large cyclic entropy. With degree-based entropy, the variation of degree entropy for a real network is closer to the scale-free network with varying network sizes.

In [20], by applying certain cyclic optimization algorithm in the case of a directed cyclic network, all the networks evolved to same optimal type which is a random network although the networks varied in size. Therefore, the random network is said to achieve the best equilibrium state in social networks.

10.2.7 Navigation Entropy

Navigation entropy [17] is used to measure complexity or randomness of a complex network. Navigability determines possible paths between a pair of nodes and thus models the structure of a network. It gives information about shortest routing paths. Navigation entropy helps in determining a routing scheme which establishes shortest routing paths. A random walk is performed from each node and entropy is computed for each node. Average entropy across all nodes is calculated to reveal network complexity. Navigation entropy is given below.

$$H(k_i) = -\sum_j p(k_{ij}) \log p(k_{ij}) \tag{10.18}$$

$$H(G) = \frac{\sum_i H(k_i)}{N} \tag{10.19}$$

where $H(k_i)$ is navigation entropy of the ith node, and $H(G)$ is average navigation entropy of the network G with N nodes. In Equation (10.18), $p(k_{ij})$ is the probability that node i is linked to node j and can be obtained as follows:

$$p(k_{ij}) = \frac{A_{ij}}{\delta_i} \tag{10.20}$$

where $\delta_i = \sum_j A_{ij}$, is a possible number of out-links for a given node i and A is considered an $N \times N$ adjacency matrix of a graph G with N nodes where $A_{ij} = 1$ if there is a link between i and j; otherwise $A_{ij} = 0$. If we take $A^{(0)} = A$ as the original adjacent matrix, the ith step of reachability matrix is $A^{(i)} = A^{(i-1)} \times A$. Navigation entropy of the i_{th} step of reachability matrix is given below.

$$H^i(G) = H(A^{(i)}) \tag{10.21}$$

A node i has δ_i choices to route a message to the next node. Using a deterministic routing scheme, one of the out-links is chosen and one-step entropy of the network is estimated using Equation (10.21). In a random routing scheme, $A^{(k)}$ and $\delta^{(k)}$ represent the reachability and degree of the ith node after $k-1$ steps, where $A^{(k)}(i,j)$ represents the number of paths from node i to node j at the kth step. Therefore, navigation entropy is maximum when there are more choices or more uncertainties in choosing the outgoing link to the next node. Thus, a fully connected graph has the maximum navigation entropy.

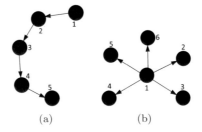

FIGURE 10.7: Navigation entropy: (a) line network and (b) star network

In the directed line network shown in Figure 10.7(a), each node has only one outgoing and one incoming link; therefore $H(G) = 0$ using Equations (10.18), (10.19), and (10.20). Thus, no information is needed to route the message to the next node since only one outgoing link is available. In Figure 10.7(b), node 1 has six outgoing links, which makes $H(G) = H(1) = 0.298$.

A directed line or ring network has the smallest navigation entropy $H(G)$ since the nodes have only one possible path for exchanging information. This property makes a ring network highly structured. Navigation entropy is maximum where a node reaches any other node in the graph with equal probability. This method is used for structure analysis of real-world networks.

10.2.8   Entropy Based on Automorphism Partition

Degree-based entropy measurement is successful in measuring the heterogeneity of a network. However, it cannot differentiate between nodes with the same degree. In [16], automorphism partition is done to differentiate nodes with equal degrees. Nodes in each partition group have the same degree, clustering coefficient, number of cycles which pass through the node and length of the longest path starting from the node. Thus, automorphism partition is a finer partition method than the degree-based approach where the nodes are only differentiated with respect to degree. Entropy based on automorphism partition is given as follows:

$$H = - \sum_{1 \leqslant k \leqslant P} p(k) \log p(k) \tag{10.22}$$

where $p(k)$ is the probability that the vertex belongs to V_k of the automorphism partition $P = \{V_1, V_2, \ldots, V_k\}$ and $p(k)$ is calculated using the following equation.

$$p(k) = \frac{|V_k|}{N} \tag{10.23}$$

where $|V_k|$ represents the number of nodes in the kth partition and N is the total number of nodes in the network.

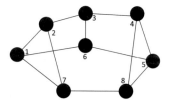

FIGURE 10.8: Entropy based on automorphism partition [16]

In Figure 10.8, automorphism partition of the network is shown as $P = \{3,6\},\{7,8\},\{1,2,4,5\}$, such that nodes in each partition are structurally equivalent. Entropy of the network is equal to 1.039 when using Equation (10.22) and Equation (10.23), whereas, $H_{max} = \log N = 2.079$. Maximum entropy corresponds to asymmetric networks and minimum entropy is found in transitive networks. Transitive networks are entirely homogeneous structure.

Various real-world networks were studied by measuring entropy based on automorphism partition and a comparison made with the degree-based entropy in [16]. The automorphism partition calculated heterogeneity and complexity of a network more effectively and accurately.

10.2.9 Entropy as Measure of Network Evolution

10.2.9.1 Degree-Based Entropy as Measure of Network Evolution

Entropy analysis has been carried out to determine changes in node and link connections to understand the evolution of small-world networks in a ring structure [9]. Experiments were done using the Watts and Strogatz (WS) network model [3] to understand variations in network entropy. To understand nodal changes, degree-based entropy definitions were the same as described in Equation (10.1) and Equation (10.2). Entropy increases with increase in disorder and becomes maximum when all probabilities are equal.

To analyze entropy with respect to changes in link connections, edges in a ring structure were randomly rewired with probability p and changes in network entropy were observed for each iteration by varying p and varying the number of neighbor nodes. With increasing value of p, network entropy followed a U-shaped curve in the course of generation processes and changes in numbers of neighbors affect the rise time of the curve.

10.2.9.2 Robustness-Based Entropy as Measure of Network Evolution

Network evolution was studied using entropy as a measure by applying Darwinian principles of variation and preferential selection [21]. The addition of a new node to an ancestral network generated an ensemble of networks with entropies $H_{min} = H_1 \leqslant H_2 \ldots \leqslant H_N = H_{max}$ such that:

$$\delta(H_i) \equiv \frac{H_i - H_{min}}{H_{max} - H_{min}}; 0 \leqslant \delta(H_i) \leqslant 1 \qquad (10.24)$$

In Equation (10.24), $\delta(H_i)$ is the normalized entropy value of the ith network in the ensemble and H is defined based upon its robustness [21].

Selection decides which network is chosen among the ensemble. The probability of selecting the network is given as:

$$\Pi(H_i) \propto \begin{cases} \delta(H_i)^T & \text{for } T \geqslant 0 \\ (1 - \delta(H_i))^{-T} & \text{for } T < 0 \end{cases} \qquad (10.25)$$

where T is a free parameter depending on environmental conditions.

With a negative value of T, the network evolved to a regular network. For $T = 0$, a random network was obtained. For $T \geq 0$, a scale-free network was obtained but as the network size increased beyond a certain value, the network evolved to become a star network [21].

10.3 Link Length Variability Entropy

In Section 10.2, we studied entropy as a measure of degree distribution, changing topology configuration, connectivity centrality of node clusters, number of cycles and loops, node navigability, and automorphism partition. The physical separation of neighboring nodes affects the time and cost of information transmission and it is thus necessary to study link length properties in networks.

The length of a link is an important parameter in modeling. For example, in wireless mesh networks, link length determines the spectral reuse efficiency of a network. Single-hop systems provide lower reuse efficiency than multi-hop networks that involve multiple nodes along a path from source to destination. Distribution of link lengths varies among network types.

We propose a new metric called link length variability entropy (LLVE) to measure the variations in link lengths utilized in various network types. Regular grid networks, small-world networks, ER networks, scale-free networks, and spatial wireless networks were analyzed. Entropy based on LLVE is calculated below.

$$H = -\sum_{i \neq j} p_{ij} \log p_{ij} \tag{10.26}$$

In Equation (10.26), p_{ij} is the probability of variation of length of a link connecting two adjacent pairs of nodes i and j as shown below.

$$p_{ij} = \frac{|L_{avg} - D_{ij}|}{\sum_{i \neq j} |L_{avg} - D_{ij}|} \tag{10.27}$$

In Equation (10.27), L_{avg} is average link-length of the network. D_{ij} is the length of each link connecting two adjacent nodes i and j. Figure 10.9 shows the entropy variation among the network types.

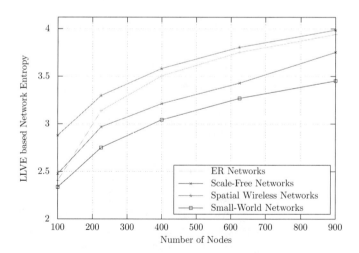

FIGURE 10.9: Entropy based on link length variability

The regular grid network showed the minimum entropy equal to zero. All links were of equal length, showing no variation. Maximum variability was found in the spatial wireless network. The order of entropy was *regular network<*

small-world network < scale-free network < ER network < spatial wireless network. In spatial networks, since the information exchange between the nodes is limited to a certain range, the lengths of most of the links are same which means most of them have equal variations in link length probability as given in Equation (10.25). Thus, entropy of spatial wireless network is higher than entropy of other networks.

Small-world networks performed similarly to the regular network since the small-world network has a few nodes that create long-links with other nodes in a usual grid network. Therefore, the presence of long-links showed a slight deviation in entropy from that of regular grid network. A regular grid network has minimum entropy because there is no variation in link lengths between any pair of nodes in the network.

10.4   Comparison of Various Entropy Definitions

The characteristic, maximum and minimum entropy network, and time complexity for estimating entropy are compared in Table 10.2 for the definitions discussed in Section 10.2.

10.5   Conclusion

Entropy is useful to measure randomness in any system. Any changes in a system over space or time will change its entropy. In this chapter, various entropy models have been tested via Shannon's entropy definition to understand different network properties. This chapter also discusses the time complexity involved in the calculation of each entropy model.

Attempts to measure the robustness, heterogeneity, connectivity, centrality, and other network properties were covered by different entropy models; however, more work must be done to understand the evolution of networks.

Link length variability entropy (LLVE) reveals behaviors of different networks based on the lengths of links. Length of a link affects communication between two nodes. Thus, the new probability metric defined for LLVE discriminates networks based on variations in link length. Among the network models studied in this chapter, regular grid networks showed minimum entropy and spatial wireless networks showed maximum entropy. Thus, spatial wireless networks showed more variation in link lengths than regular grid networks that showed no variation.

TABLE 10.2: Qualitative Comparison of Various Entropy Definitions

Entropy Type	Definition	Characteristic	Maximum Entropy	Minimum Entropy	Time Complexity
Nodal Degree	$H = -\sum_k p(k) \ln p(k);\ p(k) = \frac{d(k)}{\sum_{k=1}^{N} d(k)}$	Heterogeneity	Homogeneous	Heterogeneous	$\mathcal{O}(N)$
Topology	$H(N,p) = -\sum_{k=0}^{M} C_k^M p(k) \log_2 p(k);\ p(k) = p^k(1-p)^{M-k};\ M = \frac{N(N-1)}{2}$	Dynamics	At $p = 0.5$	At $p = 0, 1$	$\mathcal{O}(N^4)$
Connectivity	$H = -\sum_{k=1}^{n} p(k) \ln p(k);\ p(k) = \frac{n_k}{N}$	Disconnected network reliability	Network with all isolated nodes	Connected network	$\mathcal{O}(N)$
Centrality	$H = -\sum_{i=1}^{n} p(k_i) log_2 p(k_i);\ p(k_i) = \frac{spaths(k_i)}{spaths(k_1,\ k_2,\ k_3,....,\ k_M)}$	Finding central node	Symmetric networks	Nodes with unequal probability distribution of links	$\mathcal{O}(N^2 \log N)$
Clustering	$H(k) = -p_i(k) \log_2 p_i(k) - p_o(k) \log_2 p_o(k);\ H(G) = \sum_k H(k);\ p_i(k) = \frac{n}{N(k)};\ p_o(k) = 1 - p_i(k)$	Cluster quality	Fewest intra-connection clusters	Most intra-connection clusters	$\mathcal{O}(N)$
Cyclic	$H = -\sum_k p(k) \log p(k);\ p(k) = $ fraction of cycles of length k	Social network classification	Small-world	Scale-free	$\mathcal{O}(N)$
Navigation	$H(k_i) = -\sum_j p(k_{ij}) \log p(k_{ij});\ H(G) = \frac{\sum_i H(k_i)}{N};\ p(k_{ij}) = \frac{M_{ij}}{\delta_i};\ \delta_i = \sum_j M_{ij}$	Structural analysis	Star	Line	$\mathcal{O}(N)$
Automorphism partition-based	$H = -\sum_{1 \leqslant k \leqslant P} p(k) \log p(k);\ p(k) = \frac{|V_k|}{N}$	Differentiating vertices of same degrees	Asymmetric	Transitive	$\mathcal{O}(N)$
Link length variability	$H = -\sum_{i \neq j} p_{ij} \log p_{ij};\ p_{ij} = \frac{|L_{avg} - D_{ij}|}{\sum_{i \neq j} |L_{avg} - D_{ij}|}$	Link length behavior for classifying network	Spatial wireless	Regular grid	$\mathcal{O}(N)$

Bibliography

[1] P. Erdős and A. Rényi, On the evolution of random graphs, *Publication of the Mathematical Institute of the Hungarian Academy of Science*, Vol. 5, pp. 17–61, 1960.

[2] X. F. Wang and G. Chen, Complex networks: small-world, scale-free and beyond, *IEEE Circuits and Systems Magazine*, Vol. 3, pp. 6–20, September 2003.

[3] D. J. Watts and S. H. Strogatz, Collective dynamics of small-world networks, *Nature*, Vol. 393, pp. 440–442, June 1998.

[4] A.-L. Barabási and R. Albert, Emergence of scaling in random networks, *Science*, Vol. 286, pp. 509–512, October 1999.

[5] T. M. Cover and J. A. Thomas, Elements of information theory, *John Wiley & Sons*, 2012.

[6] G. Bianconi, The entropy of randomized network ensembles, *Europhysics Letters*, Vol. 81, p. 28005, January 2008.

[7] J. Wu, Y.-J. Tan, H.-Z. Deng, and D.-Z. Zhu, Heterogeneity of scale-free networks, *Systems Engineering-Theory & Practice*, Vol. 27, pp. 101–105, May 2007.

[8] B. Wang, H. Tang, C. Guo, and Z. Xiu, Entropy optimization of scale-free networks robustness to random failures, *Physica A: Statistical Mechanics and its Applications*, Vol. 363, pp. 591–596, May 2006.

[9] L. Shuo, Y. Chen, Z. Zhang, and W. Li, The analysis for small-world network's evolution based on network entropy, *AASRI Procedia*, Vol. 5, pp. 274–280, November 2013.

[10] L. Ji, W. Bing-Hong, W. Wen-Xu, and Z. Tao, Network entropy based on topology configuration and its computation to random networks, *Chinese Physics Letters*, Vol. 25, p. 4177, November 2008.

[11] L. Wu, Q. Tan, and Y. Zhang, Network connectivity entropy and its application on network connectivity reliability, *Physica A: Statistical Mechanics and its Applications*, Vol. 392, pp. 5536–5541, November 2013.

[12] A. Abraham, A.-E. Hassanien, and V. Snasel, Computational social network analysis, *Computational Social Network Analysis, Computer Communications and Networks*, Vol. 1, 2010.

[13] E. C. Kenley and Y.-R. Cho, Entropy-based graph clustering: Application to biological and social networks, in *Proceedings of IEEE International Conference on Data Mining*, pp. 1116–1121, December 2011.

[14] K. Mahdi, M. Safar, I. Sorkhoh, and A. Kassem, Cycle-based versus degree-based classification of social networks, *Journal of Digital Information Management*, Vol. 7, pp. 383–389, 2009.

[15] I. Sorkhoh, K. Mahdi, and M. Safar, Cyclic entropy of complex networks, in *Proceedings of IEEE/ACM International Conference on Advances in Social Networks Analysis and Mining*, pp. 1050–1055, August 2012.

[16] Y.-H. Xiao, W.-T. Wu, H. Wang, M. Xiong, and W. Wang, Symmetry-based structure entropy of complex networks, *Physica A: Statistical Mechanics and its Applications*, Vol. 387, pp. 2611–2619, April 2008.

[17] X. Sun, Evaluating structure of complex networks by navigation entropy, in *Proceedings of IEEE International Conference on Semantics, Knowledge and Grids*, pp. 229–232, October 2012.

[18] R. Cohen, K. Erez, D. Ben-Avraham, and S. Havlin, Resilience of the Internet to random breakdowns, *Physical Review Letters*, Vol. 85, p. 4626, November 2000.

[19] R. Timo, K. Blackmore, and L. Hanlen, On entropy measures for dynamic network topologies: Limits to MANET, in *Proceedings of IEEE Australian Workshop on Communications Theory*, pp. 95–101, February 2005.

[20] N. El-Sayed, K. Mahdi, and M. Safar, Cyclic entropy optimization of social networks using an evolutionary algorithm, in *Proceedings of IEEE International Conference on Computational Science and Its Applications*, pp. 9–16, July 2009.

[21] L. Demetrius and T. Manke, Robustness and network evolution– an entropic principle, *Physica A: Statistical Mechanics and its Applications*, Vol. 346, pp. 682–696, February 2005.

Chapter 11

Intelligent Technique for Clustering and Data Acquisition in Vehicular Sensor Networks

Amit Dua

Thapar University

Neeraj Kumar

Thapar University

Seema Bawa

Thapar University

11.1 Introduction

In recent decades, communication technology has undergone a revolution that changed lifestyles all over the world. Far-flung people can share information via the Internet by using several types of personal computers, smart phones, and other devices that are connected by following rules and regulations known as protocols.

Protocols are specific to the types of networks they control; the networks are classified broadly as wired and wireless. Wired networks have fixed infrastructures containing dedicated servers, switches, and routers to support various user activities. Wireless networks generally do not have fixed infrastructures. Their mobile clients are connected by mobile routers and access points (APs) [1-9].

Technology users are highly mobile and they require seamless connectivity from wired and wireless networks. They demand fast Internet connections, the ability to communicate wherever they are, and the ability to find immediate help in case of an emergency. Wireless networks are generally classified according to parameters such as node mobility, topology management, and applications. The main classes are:

- Mobile ad hoc networks (MANETs)

- Wireless sensor networks (WSNs)

- Vehicular ad hoc networks (VANETs)

VANETs have gained popularity because of their flexibility that allows them to provide uninterrupted services to mobile users. Academia and industry have developed many applications for VANETs [10-13] to ensure safety and comfort of passengers.

Vehicles are now equipped with sensors, software, and other equipment that allows passengers to transfer receive data via remote keyless devices, personal digital assistants (PDAs), laptop computers, and mobile phones. The integration of communication technology in vehicles greatly enhances passenger safety. Various sensors can assess conditions inside and outside vehicles. In another area, sensors can receive and transmit health data from a human body and transmit it to a remote location for monitoring and treatment. For example, VANETs can be combined with modern diagnostic systems known as body sensor networks (BSNs) to send vital signs of a patient to a hospital.

VANETs equipped with sensor nodes can handle critical applications such as generating alarms to police and fire stations in case of emergency. Houses are equipped with APs to transfer and receive data and generate alarms as required. Similarly, data collected by vehicles can be used to find optimized routing in remote areas.

Short- or long-range techniques can achieve vehicle-to-sensor communications. Short-range techniques such as Bluetooth, infrared, and Zigbee can transmit data between vehicles and sensors. Bluetooth works with a 10-meter range and can transfer data at a rate of 2 megabits per second (Mbps). Infrared is a line-of-sight system. Wavelengths range from 700 nanometers to 1 millimeter and maximum transmission distance is 10 meters. Zigbee uses less power for communication and thus features extended battery life.

Long-range systems include 4G broadband, WiFi, worldwide interoperability for microwave access (WiMAX), and long-term evolution (LTE). 4G and WiMAX follow IEEE 802.16e and 802.16m standards for communication. LTE can provide 100 Mbps downstream and 30 Mbps upstream data rates. It can follow IEEE 802.11a, b, g, and p standards that work at the band between 2.4 and 5.9 GHz [14-16].

VANETs provide two types of communications: vehicle-to-vehicle (V2V) and vehicle-to-infrastructure (V2I). V2V is similar to peer-to-peer (P2P) communication; it handles communications among vehicles on a road. In V2V, information is shared among all vehicles in a range without need to communicate with an infrastructure.

Vehicles in a V2I network use the nearest APs, generally known as road side units (RSUs) that provide seamless connectivity to all vehicles on a road. RSUs are installed at optimum locations along roads to maintain consistent coverage of vehicles [10-13]. Figure 11.1 shows the general VANET scenario including RSU locations and vehicle movements.

VANETs have been highly successful due to advances in the automotive and wireless technology fields. One reason behind VANET growth was to ensure safety of passengers in cases of road emergencies. Furthermore, individuals want the ability to use handheld devices for remote control from any location and VANETs meet this need.

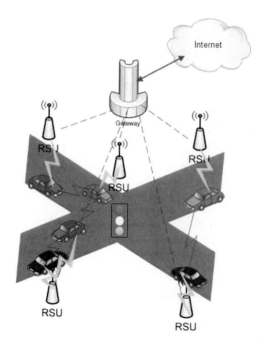

FIGURE 11.1: Schematic diagram of vehicles and interconnection between vehicles

VANETs have been incorporated into intelligent transport systems (ITSs). A VANET can act as a "computer on wheels" or a "network on wheels" and works as an ad hoc network between and among vehicles and RSUs. A VANET utilizes two types of nodes: mobile on-board units (OBUs) and static RSUs. Communication researchers are working on new VANETs concepts to provide the following benefits for end users:

- Increase traveler safety

- Increase traveler mobility

- Decrease traveling time

- Protect the environment and conserve energy

- Efficiently expand transportation systems

- Increase travel luxuries

Several challenges must be addressed by the research community working in this domain. The biggest challenge for VANETs is the mobility of nodes that creates topological changes. Secure message transmission under changing mobility conditions is another concern. Both issues must be resolved by new systems designed to increase transportation efficiency and road safety.

Packet delivery ratio (PDR) is low at present because there are not enough vehicles to transmit information. As velocity and density increase, the PDR will increase gradually to a maximum density of 300 and velocity of 70 kilometers per hour. Beyond those points, the system will be impacted by excessive packet exchanges caused by density and unstable links resulting from high mobility of vehicles. Figure 11.2 shows PDR variations caused by velocity and density.

For a vehicle carrying many sensor units, the maximum occurs at 1000 sensors and 80 kilometers per hour velocity. The variation is shown in Figure 11.3.

11.1.1 Organization

The rest of this chapter is organized as follows. Section 11.2 describes various mobility models used for data collection in VANETs. Clustering with existing algorithms is discussed in Section 11.3. Section 11.4 gives a description of data collection and mobility techniques. Section 11.5 covers the issues in VANETs and discusses differences and similarities between VANET and mobile ad hoc networks in Section 11.6. Section 11.7 discusses the various applications of VANETs. Section 11.8 of this chapter discusses various simulation tools and their comparison. Finally, Section 11.9 concludes the chapter.

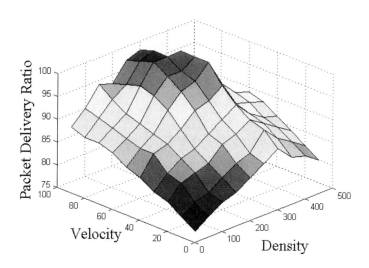

FIGURE 11.2: Variation in packet delivery ratio with increasing velocity and density

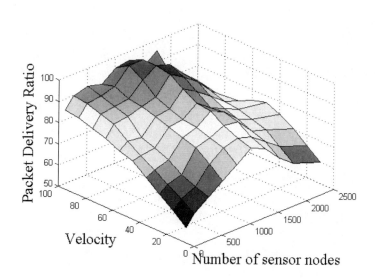

FIGURE 11.3: Variation in packet delivery ratio with increasing velocity and number of sensors

11.2   Mobility Models for Data Collection

The movement patterns of mobile nodes (variation in location, velocity, and acceleration over time) were examined in mobility models. These models play an important role in determining protocol efficiency and the appropriate model should be chosen according to the scenario requirements. A few important models are described in this section and Figure 11.4 shows their classifications.

Random Waypoint Mobility Model: Johnson and Maltz [19] proposed this model. The mobile node moves in a specific direction for a certain period, then pauses for a limited time. After a pause, the node starts moving in a random direction at a random speed between minimum and maximum permitted speeds (V_{min} and V_{max}, respectively). After moving in the new direction for a certain time, the node pauses again and the process is repeated.

Random Walk Model: Nodes change their directions and speeds after moving for fixed periods. The new direction $\theta(t)$ can be anything from 0 to 2π. The new speed can follow Gaussian or uniform distribution. There is no pause between changes of velocity and direction; this is the only difference from the random waypoint mobility model.

Both models are easy to understand and implement and both have limitations. For example, they do not consider previous or current velocities in deciding new velocity, so acceleration, stops, and turns are sudden. This limitation is called temporal dependency of velocity.

Gauss-Markov Mobility Model: The temporal dependency of velocity is overcome by this model as current velocity is dependent on previous velocity. The velocity is modeled as Gauss-Markov stochastic process and is assumed to be correlated over time. For a two-dimensional simulation field,

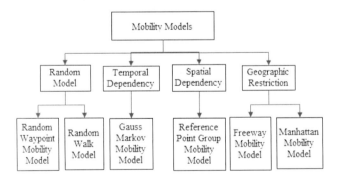

FIGURE 11.4: Classification of various mobility models

TABLE 11.1: Relative Comparison of Mobility Models

Mobility Model	Temporal Dependency?	Spatial Dependency?	Geographic Restriction?
Random Waypoint	No	No	No
Reference point group	No	Yes	No
Freeway	Yes	Yes	Yes
Manhattan	Yes	No	Yes

Gauss-Markov stochastic process is represented as:

$$v_t^x = \alpha v_{t-1}^x + (1 - \alpha)v^x + \sigma^x \sqrt{1 - \alpha^2} w_{t-1}^x \qquad (11.1)$$

$$v_t^y = \alpha v_{t-1}^y + (1 - \alpha)v^y + \sigma^y \sqrt{1 - \alpha^2} w_{t-1}^y \qquad (11.2)$$

α is the parameter reflecting randomness of Gauss-Markov process. If $\alpha = 0$, the model is memory less and the equations derived show that it is the same as the random walk model.

In mobility models discussed to this point, location, speed, and movement direction are not affected by nodes in the neighborhood. The reference point group mobility model takes these factors into consideration.

Reference Point Group Mobility Model: This model contains group leader nodes and group member nodes. Movements of the entire group are determined by the movements of the group leader. This quality is useful in limiting the speed of vehicles on freeways to prevent collisions and also on battlefields and in disaster relief situations in which team members must follow a group leader.

The nodes in the mobility models described above can move theoretically in any direction although in reality their movements are constrained by buildings, road conditions, and other types of obstacles. An example of a model incorporating geographic restrictions and environmental conditions is the Manhattan mobility model.

Manhattan Mobility Model: Mobile nodes travel roads in horizontal and vertical directions based on geography. The Manhattan model follows a probabilistic approach in selecting a direction at an intersection. A vehicle can turn left or right or continue in the same direction based on certain probabilities. Table 11.1 compares temporal dependency, spatial dependency, and geographic restriction characteristics of various mobility models.

11.3 Clustering

Two methods implement communications between vehicular networks and the Internet backbone. The first is direct communication between mobile nodes and the Internet. The second method employs a wireless WAN and wireless LAN in a cluster structure.

The first method views a vehicular network as a flat topology in which each node is equal. The WAN and LAN work in parallel. Communication can proceed with or without the LAN. The second utilizes a cluster structure to generate hierarchical vehicular networks.

Existing Clustering Algorithms for VANETs: A VANET is a special type of a MANET and an important component of intelligent transportation systems. A VANET provides efficient communication about speed, direction, and acceleration changes and real-time traffic information like safety alert messages to vehicles in the network.

The clustering strategy for data transmission involves selection of cluster head for each cluster. The cluster head is responsible for relaying data and control packets within and outside the cluster. The approach was initially proposed for MANETs and later found useful in VANETs. The systems differ in that VANET nodes have greater mobility and thus face more frequent topology changes. Other factors are the constraints of road conditions that affect vehicle trajectory and driver behavior, for example, speed changes involved in overtaking other vehicles. Drivers can easily acquire speed, direction, and location based on GPS and similar systems.

Some organization requirements for MANET clustering do not apply to VANETs. For example, energy efficiency is critical for MANETs but not relevant for VANETs since the vehicles in which they are installed recharge their batteries during journeys. This means that traditional clustering algorithms used for MANETs cannot be applied directly to VANETs. These differences should be considered when designing clustering strategies for VANETs.

Research is ongoing to ensure that VANET clusters remain stable and provide reliable and efficient communications. Several techniques are under study to ensure stable clustering involving signal strength, positioning of nodes from cluster heads, and the velocities, directions, and destinations of nodes. The proposals based on clustering algorithms are described below.

Clustering-Based MAC Protocols: Yvonne et al. introduced a clustering scheme for medium access control (MAC) to ensure fairer medium access and reduce the effects of the hidden station problem. Each node maintains two tables to track changes in topology. The first is for neighboring nodes and the second is for adjacent clusters. Each node sends data only in its own time slot. The slots are allotted according to the amount of data the nodes want to transmit.

The authors simulated the proposed protocol with fixed parameters such as TDMA frames and lots and evaluated functionally. The results showed good cluster stability and data transmission in several scenarios involving varying traffic densities. The protocol worked well in low and medium traffic densities but achieved low data communication rates in rush hour traffic.

11.4   Data Collection with Mobility in VANETs

It is challenging to collect real-time data for processing and analysis in VANETs due to high mobility. Following are the methods used for collecting data in VANETs [14].

- Triangulation method

- Vehicle re-identification

- GPS-based method

- Sensor-based methods

11.5   Issues in VANETs

Although several issues influence the applications of VANETs, the main issues having direct impact on the performance of any solution are described below.

11.5.1   Mobility

The mobility of vehicles depends on the density on the roads. The higher density roads have less mobility. Density may vary according to time of the day. The rush hours in mornings and evenings have more density and impact mobility on roads. The weekends involve fewer vehicles on roads and mobility is high.

Different mobility models can be used, i.e., Manhattan, freeway, and others. In cities where the roads are straight and intersect at right angles, the most suitable mobility model is the Manhattan. Some models treat movement of vehicles as random and any vehicle can take any path. These vehicles follow the freeway mobility model. There is no restriction on any vehicle movement in this model.

Any mobility model must consider two scenarios, i.e., highway and city. In the highway scenario, vehicles are sparsely distributed and speed is very high. While in the city scenario, the vehicles are densely distributed and their velocity is low.

11.5.2   Scalability

Solutions for small scenarios might not work on larger scenarios. Various constraints must be considered before extending a smaller model to be used in a larger network. First of all the density and mobility of vehicles might not follow the same pattern. The weather conditions and behaviors of drivers are the factors that need to be incorporated before extending the model. In smaller networks, the packet exchanges between vehicles are limited and exchanges increase exponentially as the number of vehicles increases. Problems like the broadcast storm problem can emerge because of higher collision rates and contention of packets as we increase the scalability of network. These factors must be considered when increasing scalability.

11.5.3   Channel Utilization

Cognitive radio-based scenarios must be used in VANETs. One vehicle acts as the primary user and another acts a secondary user. When the primary user is accessing the channel, the secondary user cannot access it and vice versa. If both users attempt to use the channel at the same time, a contention resolution policy must be used. The multiple channels enhance battery life and improve bandwidth usage, thus improving packet delivery ratio, channel utilization, and throughput. As more vehicles use this scheme, allotments of primary vehicle status and channel allocation are handled on a priority basis.

11.5.4   Security and Privacy

The public key infrastructure (PKI) system is used to ensure vehicle security. A centralized key distribution center (KDC) distributes keys based on vehicle identification data. System vehicles must register with the KDC before they use the system. The KDC generates keys and distributes them to vehicles. If a key is lost or misplaced, it is revoked.

If a vehicle changes its domain, MIPV6 is used to ensure a secure change. The home domain is where the vehicle is registered; if a vehicle goes beyond its home domain, it is moving in a foreign domain. Time stamping ensures security. A conflict resolution list (CRL) identifying misbehaving vehicles is updated after every time stamp and the next time stamp revokes misbehaving nodes.

11.5.5   Inefficient Management of Data

Data management is the biggest challenge in VANETs because of varying traffic densities at many locations and at different times and days of the week.

Data management depends on the deployment sites and intensities of sensors and also on vehicle locations. Only cloud computing provides centralized locations for storing sensed data from vehicles. Even the P2P systems encounter problems when a vehicle moves out of range. Data management systems must be capable of predicting movements and preventing accidents.

11.6 Similarities and Differences of VANETs and MANETs

VANETS are special mobile ad hoc networks (MANETs). Both use deploy nodes in an ad hoc manner. Nodes in MANET can directly communicate with each other without centralized authority or infrastructure. The major benefit of MANETs is fast deployment of nodes and network setup on battlefields and sites of natural calamities.

11.6.1 Comparison of Characteristics

There are some unique properties of VANETs that make them unique. Because of these properties, VANETs encounter considerable challenges and need exclusively designed protocols. Some of the characteristics that make them unique are discussed next.

As the velocity can go as high as 150 kilometers per hour, there are frequent changes in topology. This results in limited time for communication between vehicles. This time is even smaller if the vehicles are moving in opposite directions. The transmission range can be increased to counteract the decreased communication time but this would result in more collisions of packets. The higher contention rate will degrade the throughput of the system. To improve the efficiency, low latency protocols are required.

To counteract high mobility and deliver information on time, broadcasting messages can be one solution. Despite high vehicle speeds and short reaction times, messages must be delivered in fraction of seconds. Communication urgency is an intrinsic property of VANETs and not a requirement for MANETs.

MANETs can work with predictive routing schemes; the schemes have not been successful in VANETs because of rapid and unpredictable location changes. An efficient routing table in VANET would become obsolete immediately or would require amounts of channel utility that would degrade network efficiency.

The topology of VANETs changes as vehicles move along roads. The changes are predictable as vehicles move only on fixed roads.

Minimizing the energy use to maximize life is a major objective of a MANET. Energy use is not an issue for VANETs because vehicle batteries are charged during use and can last for months and even years. In addition to monitoring and transmitting vehicle location and other data, vehicle batteries are capable of powering accessories such as air conditioners and music systems; power is always an issue of concern in MANETs.

VANETs were deployed initially on small numbers of vehicles. Increased numbers of transceivers installed in vehicles will lead to frequent fragmentation of a network. Fragmentation must be considered in the design of VANET protocol.

User privacy and security must be ensured for VANET technology to be widely accepted. A moving vehicle is considered a private space. Monitoring vehicle activities is a breach of personal privacy and not acceptable even if implemented by authorities. A malicious third party user could use such information for illicit purposes and cause harm to legitimate network users.

Another misuse of a VANET is raising a false alarm by tampering with propagated messages. This misuse is a criminal matter and should be dealt with accordingly. The main characteristics of VANETs are as follows:

- Mobility of nodes is very high

- The topology can be predicted

- Latency requirement is critical for safety applications

- Power is no issue

- Fragmentation possibility is high

- Migration rage is slow

- Security and privacy are critical factors

Centralized infrastructure connected to the Internet can be deployed on roads. However, because the numbers and sizes of roads continue to increase, installment of these devices on all roads is impossible. Road side units can be deployed at regular intervals. They can efficiently communicate with fast moving vehicles, increase communication rates, and decrease latency. They do not require charging or complex centralized infrastructures and thus make VANET application a reality. VANETs are special versions of MANETs that use fixed road side units to communicate data to and from in-vehicle devices.

11.7   VANETs Applications

More than 100 safety-related and non-safety related applications have been found for VANETs. Some of them are listed below.

Co-operative Collision Warning: Messages warning of collision are broadcast to nearby vehicles. This can save driver time and ensure safety of nearby vehicles and passengers.

Lane Changing Warning: Warning can be issued to nearby vehicles whenever a driver abruptly changes vehicle speed and moves to another lane.

Intersection Collision Warning: These warnings are generated by the road side units, not by vehicles. They notify approaching vehicles of collisions to enable them to take preemptive actions such as applying brakes or changing direction.

Approaching Emergency Vehicle: The approach of a high speed vehicle such as an ambulance or other priority vehicle must be broadcast to other vehicles in the vicinity.

Work Zone Warning: Whenever construction or other maintenance work is performed on the roads, vehicles must be informed so they can change the route. This saves time and prevents traffic jams.

Inter-Vehicle Communications: Vehicles can communicate amongst themselves directly using onboard units or through road side units.

Electronic Toll Collection (ETC): On board units are charged as soon as they pass through a toll collecting RSU. The charge is deducted through a centralized infrastructure that enters the required amount.

Parking Lot Payment: Similar to toll collection, parking lot payment can be made by a toll collecting infrastructure that deducts the amount from pre-charged on board units.

Traffic Management: Real-time traffic monitoring can be performed by interactions of on board units with road side units. This can be used to select an appropriate path having minimum traffic. In addition, in case of an accident traffic can be managed properly.

For evaluating the performance of protocols it is not possible to deploy vehicles with AUs and RSUs in real scenarios. Simulations of environments similar to real scenarios are performed. For carrying out simulations in VANETs the following main tools are used.

11.8   Simulation Tools

OPNET [20]: OPNET provides libraries and graphical packages for simulation scenarios of MANETs, satellite networks, Bluetooth, WiMAX and IEEE 802.11 wireless LANs. Graphical editor interface can be used for physical layer to application process modulation. OPNET development has three basic phases:

1. To configure node models

2. To set up connections between nodes and to form the network

3. To specify the parameters on which simulation is to be performed

There are three main files required in OPNET for simulation namely network configuration file, node configuration file and global parameter file. Each has its own specific purposes.

GloMoSim and QualNET [21][22]: GloMoSim is a network simulator and generates traces for random waypoint-like mobility models. Though GloMoSim was discontinued in 2000, its commercial version QualNET, which is a discrete event simulator with efficient parallel kernel, was launched. Its computationally efficient code allows simulations of 5000 nodes very quickly. Specific modules like animator, analyzer, packet tracer, and scenario designer, for handling specific functions can be incorporated. Man-in-the-loop and hardware-in-the-loop models can be simulated using QualNET in real time.

SUMO [23]: Simulation of urban mobility (SUMO) is a command line simulator that uses C++ standards and portable libraries. It is an open source microscopic level simulator which is easy to compile and can execute on any operating system. Although capable of handling different networks and types of vehicles, it can run at high speed. Many applications can be incorporated in SUMO. The graphical user interface GUISIM can be added to SUMO which is an example of its import and export capabilities.

NS-2 [24]: It is a discrete event network simulator whose core is written in C++. It has found wide popularity because it is open source. Users write TCL script specifying mobility models, topology, and other wired and wireless parameters for running a simulation. The outputs of simulation are the trace file and NAM file. Trace files record transmission, forwarding, packet drop, and packet reception events. The NAM file is a GUI visualization of the entire simulation.

J-Sim [25]: J-Sim is similar to NS-2 as it also takes TCL as input and produces event trace and animation files for use in NAM. It is also open source. The difference is that it supports different formats and is developed completely in Java. The applications in J-Sim can be tested separately as they are designed and built on an exclusive set of components. It supports the random waypoint and trajectory-based mobility models.

OMNeT++ [26]: The applications of OMNeT++ are in Internet simulation. As it is open source, it produces GUI-based output which can be plotted through GUI and has high utility. The overall design is component-based and supports new features. The support of protocols in OMNeT++ is through various modules.

VanetMobiSim [27]: VanetMobiSim is a platform coded in Java which produces mobility traces for other simulators like Qualnet, GloMoSim and NS-2 [21, 22, 24]. The models produced include motion at both microscopic and macroscopic levels. The traces produced can be for a multilane road environment with different direction flows and modelling for intersections. The mobility patterns are highly realistic and support vehicle-to-vehicle and vehicle-to-infrastructure interactions. The comparison among different simulators is shown in Table 11.2.

TABLE 11.2: Comparison of Simulators

Simulator	Type	Language Used	GUI?	Open Source?	Nodes Supported	Radio Propagation models						Multilane Support?	Strength	Weakness
						FS	TR	CS	LS	RA	RI			
NS-2	Discrete event	C++ and OTcl	No	Yes	< 100	✓	✓	✓	-	-	✓	Yes	Used for wireless and wired area.	Cannot simulate bandwidth or power consumption problem; some protocols and features are not well documented
OMNet++	Discrete event	C++	Yes	Non-commercial as well as commercial	Large number	✓	✓	✓	✓	✓	✓	Yes	Can simulate power consumption problems in WSNs	Limited number of protocols available
J-Sim	Discrete event	Java	Yes	Yes	~ 500	✓	-	-	-	-	-	-	Can simulate power consumption problem	Very long execution time
Qualnet	Discrete event	C and Java	Yes	Commercial	Up to 5000	✓	✓	✓	✓	✓	✓	-	Supports wired and wireless systems	Inadequate documentation; GUI not sufficiently detailed
VANET Mobi SIM	Discrete event and visual	Java	Yes	-	~ 500	-	-	-	-	-	-	Yes	Support for multilane roads, separate directional flows, and traffic signs at intersection	Cannot return feedback

FS = free space; TR = two-ray; CS = constant shadowing; LS = lognormal shadowing; RA = Rayleigh; RI = Ricean.

11.9 Conclusion

Vehicular sensor networks (VSNs) have been used in many applications over the past few decades. Data collection and acquisition in these networks are challenging issues which need special attention. In this chapter, we identified various issues and challenges. Various types of mobility models are described and compared using various performance evaluation metrics. The impacts of varying velocity and density on the packet delivery fraction and throughput have also been analyzed. A relative comparison of various simulation tools is also discussed in the chapter to enable the users to study the behaviors of various parameters with respect to vehicular motions.

In the future, we would like to incorporate secure data acquisition with variations in velocity and density of vehicles in traffic environments.

Bibliography

[1] A. Dua, N. Kumar, S. Bawa (2014). A systematic review on routing protocols for vehicular ad hoc networks, *Vehicular Communications*, 1(1),33-52

[2] N. Kumar, J.-H. Lee (2014). Peer-to-peer cooperative caching for data dissemination in urban vehicular communications. *IEEE Systems Journal*, 8(4), 1136-1144.

[3] N. Kumar, S. Misra, M.S. Obaidat (2015). Collaborative learning automata-based routing for rescue operations in dense urban regions using vehicular sensor networks. *IEEE Systems Journal*, 9(3), 1081-1090.

[4] N. Kumar, J.H. Lee, J.J.P.C. Rodrigues (2015). Intelligent mobile video surveillance system as a Bayesian coalition game in vehicular sensor networks: learning automata approach. *IEEE Transactions on Intelligent Transportation Systems*, 16(3), 1148-1161.

[5] N. Kumar, N. Chilamkurti, S.C. Misra (2015). Bayesian coalition game for Internet of things: an ambient intelligence approach. *IEEE Communication Magazine*, 53(1), 48-55.

[6] N. Kumar, N. Chilamkurti, J.J.P.C. Rodrigues (2014), Learning automata-based opportunistic data aggregation and forwarding scheme for alert generation in vehicular ad hoc networks. *Computer Communications*, 39 (2), 22-32.

[7] N. Kumar, S. Misra, J.J.P.C. Rodrigues, M.S. Obaidat (2014). Networks of learning automata: a performance analysis study. *IEEE Wireless Communication Magazine*, 21(6), 41-47.

[8] N. Kumar, C.C. Lin (2015). Reliable multicast as a Bayesian coalition game for a non-stationary environment in vehicular ad hoc networks: a learning automata-based approach. *International Journal of Ad Hoc and Ubiquitous Computing*, 19(3-4), 168-182.

[9] N. Kumar, R. Iqbal, S. Misra, J.J.P.C. Rodrigues (2015). Bayesian coalition game for contention-aware reliable data forwarding in vehicular mobile cloud. *Future Generation Computer Systems*, 48, 60-72.

[10] N. Kumar, S. Tyagi, D.J. Deng (2014). LA-EEHSC: learning automata-based energy efficient heterogeneous selective clustering for wireless sensor networks. *Journal of Networks and Computer Applications*, 46 (11), 264-279.

[11] N. Kumar, R. Iqbal, S. Misra, J.J.P.C. Rodrigues (2015). An intelligent approach for building a secure decentralized public key infrastructure in VANET. *Journal of Computer and System Sciences*, 81(6), 1042-1058.

[12] N. Kumar, S. Misra, J.J.P.C. Rodrigues, M.S. Obaidat (2015). Coalition games for spatio-temporal big data in internet of vehicles environment: a comparative analysis. *IEEE Internet of Things Journal*, 2(4), 310-320.

[13] N. Kumar, J.J.P.C. Rodrigues, N. Chilamkurti (2014). Bayesian coalition game as-a-service for content distribution in Internet of vehicles, *IEEE Internet of Things Journal*, 1(6), 544-555.

[14] R. Bali, N. Kumar, J.J.P.C. Rodrigues (2014). Clustering in vehicular ad hoc networks: taxonomy, challenges and solutions. *Vehicular Communications*, 1(3), 134-152.

[15] A. Dua, N. Kumar, S. Bawa, (2015). QoS-aware data dissemination for dense urban regions in vehicular ad hoc networks. *Mobile Networks and Applications*, 20(6), 773-780.

[16] N. Kumar, J. Kim, (2013). Probabilistic trust aware data replica placement strategy for online video streaming applications in vehicular delay tolerant networks. *Mathematical and Computer Modeling*, 58 (1/2), 3-14.

[17] N. Kumar, N. Chilamkurti, J.J.P.C. Rodrigues (2014). Learning automata-based opportunistic data aggregation and forwarding scheme for alert generation in vehicular ad hoc networks. *Computer Communications*, 59(1), 22-32.

[18] N. Kumar, N. Chilamkurti, J.H. Park (2013). ALCA: agent learning-based clustering algorithm in vehicular ad hoc networks. *Personal and Ubiquitous Computing*, 17(8), 1683-1692.

[19] D.B. Johnson, D.A. Maltz (1996). Dynamic Source Routing in ad hoc Wireless Networks, *Mobile Computing*, 153-181.

[20] http://www.opnet.com/

[21] X. Zeng, R. Bagrodia, M. Gerla (1998). GloMoSim: a library for parallel simulation of large-scale wireless networks. In *Proceedings of 12th Workshop on Parallel and Distributed Simulation PADS 98.* (pp. 154-161). bibitem22 http://www.qualnet.fr/

[22] http://sumo.sourceforge.net/

[23] http://www.isi.edu/nsnam/ns/

[24] http://j-sim.cs.uiuc.edu/

[25] http://inet.omnetpp.org/

[26] M. Fiore, J. Harri, F. Filali, C. Bonnet (2007). Vehicular mobility simulation for VANETs. In *Proceedings of 40th IEEE Annual Simulation Symposium (ANSS'07).* (pp. 301-309).

Index

advanced context modeler, 177
ambience monitoring agents, 177
approaching emergency vehicle, 277
APTEEN, 86
artificial neural network, 29
automated energy supply systems, 9

Bayesian belief network, 181
body node coordinator, 206

centrality entropy, 252
clustering entropy, 254
coherent-based routing, 88
competitive learning, 38
context assembler, 177
context classifier, 177
convergecast routing, 114
cooperative games, 126
coverage, 15
CStore, 177
cyclic entropy, 254

data delivery model, 16
DCLFR, 102
defuzzification, 26
degree method, 246
delay, 90
directed diffusion, 87

electronic toll collection (ETC), 277
energy expenditure, 90
entropy, 244
entropy analysis, 257
environment monitoring, 8
evolutionary computing, 48

fitness sharing, 61
flat-based routing, 85

fuzzification, 26
fuzzy logic controller, 25
fuzzy rule base, 27

game theory, 125
GEAR, 86
genetic algorithm, 49
GPSR, 86
gradient descent learning, 38

hard computing, 23
Hebbian learning, 38
HEH-BMAC, 205
hierarchical-based routing, 85
human energy harvesting, 205

independent component analysis, 183
intelligent transportation systems, 8
intersection collision warning, 277

key distribution centre, 274

LEACH, 155
learning automata, 100
linguistic states, 26
location-based routing, 86

Manhattan mobility model, 271
maximum connectivity entropy, 250
military applications, 7
mobile health, 190
multimedia WSN, 12
multiobjective genetic algorithm, 59
multiobjective optimization, 49
multipath-based routing, 87

Nash equilibrium, 127
navigation entropy, 255